Learning and Analytics in Intelligent Systems

Volume 34

The main aim of the series is to make available a publication of books in hard copy form and soft copy form on all aspects of learning, analytics and advanced intelligent systems and related technologies. The mentioned disciplines are strongly related and complement one another significantly. Thus, the series encourages cross-fertilization highlighting research and knowledge of common interest. The series allows a unified/integrated approach to themes and topics in these scientific disciplines which will result in significant cross-fertilization and research dissemination. To maximize dissemination of research results and knowledge in these disciplines, the series publishes edited books, monographs, handbooks, textbooks and conference proceedings.

Indexed by EI Compendex.

Konstantina Chrysafiadi

Fuzzy Logic-Based Software Systems

Springer

Konstantina Chrysafiadi
University of Piraeus
Piraeus, Greece

ISSN 2662-3447 ISSN 2662-3455 (electronic)
Learning and Analytics in Intelligent Systems
ISBN 978-3-031-44456-2 ISBN 978-3-031-44457-9 (eBook)
https://doi.org/10.1007/978-3-031-44457-9

This Springer imprint is published by the registered company Springer Nature Switzerland AG
The registered company address is: Gewerbestrasse 11, 6330 Cham, Switzerland

Paper in this product is recyclable.

Series Editors' Foreword

Fuzzy Logic is one of the most popular and most widely applied mathematical theories. Indeed, it was first proposed over a century ago by Łukasiewicz and Tarski [1, 2] who presented it as infinite-valued logic and called it "Multivalued Logic." The term "fuzzy" was much later introduced by the late Professor Lotfi A. Zadeh of the University of California at Berkeley in his seminal 1965 paper [3]. Since then, Fuzzy Logic has been an area of basic research and occasional controversy, but, above all, broad and particularly successful applications.

The popularity of Fuzzy Logic stems from its ability to successfully model imprecision, vagueness and uncertainty that are present in essentially all scientific and technological regimes. This characteristic feature of Fuzzy Logic resembles human representation of subjective knowledge, mimics human reasoning processes, and provides a very powerful tool to "compute" with linguistic terms.

Despite its long history, Fuzzy Logic remains an area of very active research worldwide and new applications keep on continuously emerging. Indeed, **Artificial Intelligence-empowered Software Systems** are constantly being developed and Fuzzy Logic is quite often their underlying empowering technology. As series editors of the *LEARNING AND ANALYTICS IN INTELLIGENT SYSTEMS SERIES* of Springer and to verse the scientific community in the most recent relevant developments, we are particularly happy to present the monograph at hand, by Dr. Konstantina Chrysafiadi, which is devoted to **Fuzzy Logic-based Software Systems**.

More specifically, Dr. Chrysafiadi first presents the general structure of Fuzzy Logic-based Software Systems, which consists of three basic modules, namely: (i) the Fuzzifier, (ii) the Inference Engine and (iii) the Defuzzifier. In these software systems, the Fuzzifier converts crisp numerical data into fuzzy sets. In turn, the Inference Engine produces fuzzy outputs via use of linguistic information and application of fuzzy rules over the fuzzy sets returned by the Fuzzifier. Finally, the Defuzzifier, converts the fuzzy values returned by the Inference Engine into crisp values at the system output.

Next, Dr. Chrysafiadi discusses the significant role Fuzzy Logic can play in Artificial Intelligence-empowered Software Systems, improving their efficiency and effectiveness. In this context, she presents methodologies of Artificial Intelligence which are combined with Fuzzy Logic. Finally, the book also outlines several Fuzzy Logic-based Software Systems in various disciplines, including Medicine, Education, Decision Making and Recommendation, Natural Language Processing, Automotive Engineering and Industry, Heating, Ventilation and Air-conditioning, Navigation, Scheduling, Network Traffic Management and Security.

This monograph fills in a gap in the available literature, as it verses its readers in streamline aspects and applications of Fuzzy Logic-based Software Systems. Overall, the book is very well written. It starts from introductory ideas and concepts and leads its readers to the most recent and most advanced research results in this area. Thus, the book is a valuable guide for the expert in the field of Fuzzy Logic who wishes to embed Fuzzy Logic components in Software Systems. At the same time, the book also appeals to the newcomer who wishes to learn about Fuzzy Logic and its ability to empower Software Systems. Finally, the book will certainly attract the interest of readers from other areas as well who wish to get versed in this significant scientific discipline.

As series editors, we welcome this monograph to the *LEARNING AND ANALYTICS IN INTELLIGENT SYSTEMS SERIES* of Springer and present it to the research communities worldwide. We congratulate the author for her great work, in confidence that her monograph will help its readers not only understand, but also apply the proposed Fuzzy Logic-based methodologies in various Software Systems in many and diverse scientific areas. Finally, we encourage the author to continue her research work in this important area and keep the scientific communities appropriately updated of her research results.

September 2023

Maria Virvou
Department of Informatics
University of Piraeus
Piraeus, Greece

George A. Tsihrintzis
Department of Informatics
University of Piraeus
Piraeus, Greece

Bibliography

[1] Łukasiewicz J., O logice trójwartościowej (in Polish), Ruch filozoficzny, vol. 5, pp. 170–171, 1920 (English translation: On three-valued logic, in L. Borkowski (ed.), Selected works by Jan Łukasiewicz, North–Holland, Amsterdam, 1970, pp. 87–88. ISBN 0-7204-2252-3)

[2] Łukasiewicz, J., Tarski, A., Untersuchungen über den Aussagenkalkül, *Comp. Rend. Soc. Sci. et Lettres Varsovie Cl. III*, vol. 23, pp. 30–50, 1930.

[3] Zadeh, L.A., Fuzzy sets, *Information and Control*, vol. 8 (3), pp. 338–353, 1965.

Preface

Humans several times in their everyday lives deal with situations in which data is vague, subjective and uncertain. Such data can be the security of a system, the performance of a machine, the productivity of an employee, the speed of a car, the knowledge level of a student, etc. In many circumstances, there are problems whose answer is not 'Yes' or 'No', but it can be an intermediate value, which indicates that it is partially true or false. It can be 'possibly yes', 'maybe', 'possible no', 'little', 'very much', 'less', etc. For example, in the question "is the water hot?", the answer can be 'no', or 'very little' or 'little', or 'quite' or 'very much', or 'yes'. Furthermore, there are phenomena and data that cannot be defined with certainty and accuracy. For example, a student can be 'novice', 'moderate', good', 'very good' or 'expert', concerning her/his knowledge level; the way that a vehicle moves concerning its velocity can be characterized as 'very slow' or 'slow' or 'normal' or 'fast' or 'very fast'; a person can be characterized as 'baby' or 'child' or 'young' or 'middle-aged' or 'old' according to her/his age. Therefore, Binary Logic, according to which a variable or data takes two values, the absolute true (1) and the absolute false (0), cannot be used to describe data that confront with vagueness, uncertainty and subjectivity. Such data can be described with fuzzy logic, according to which a variable or data takes all the intermediate values between 0 and 1 representing a variety of strengths of true. The main differences between Binary Logic (or Crisp Logic) and Fuzzy Logic are depicted in Table 1.

Fuzzy logic has been studied by Lukasiewicz and Tarski as infinite valued logic since 1920 [1]. However, Lotfi A. Zadeh of the University of California at Berkeley

Table 1 The main differences between Binary and Fuzzy Logic

Characteristic	Binary Logic	Fuzzy Logic
Values	0 or 1	Any value from 0 up to 1
Sets' limits	Distinct	Indistinct
Belonging degree	Absolute yes or absolute no	Absolute or partially yes or no
Mathematicalization	Boolean algebra	Fuzzy math, sets and computing with words

is considered the 'father' of fuzzy logic. He introduced the idea of fuzzy logic in 1965 as a methodology for computing with words [2]. He noticed that there are data which cannot be valued as 0 or 1 to be stored and managed by computers. Therefore, he proposed using fuzzy sets and membership functions [3], which assigned to data values which vary from 0 to 1. In such a way, he succeeded in introducing a reasoning method that imitates the human way of thinking and decision making. So, he proposed new issues, theories and operations for the calculus of logic, in order to manage effectively the data that are characterized by uncertainty, imprecision and vagueness. He proved that fuzzy logic is a generalization of the binary (Boolean) logic. As he notes in his study [2] on page 339 "the notion of a fuzzy set provides a convenient point of departure for the construction of a conceptual framework which parallels in many respects the framework used in the case of ordinary sets, but is more general than the latter and, potentially, may prove to have a much wider scope of applicability, particularly in the fields of pattern classification and information processing. Essentially, such a framework provides a natural way of dealing with problems in which the source of imprecision is the absence of sharply defined criteria of class membership rather than the presence of random variables."

The first reactions of the scientific community towards fuzzy logic were negative. Scientists believed that 'vagueness' was contrary to the basic scientific principles. Furthermore, they supported that the theory of probabilities was able to solve adequately any problem that fuzzy logic can deal with. However, later the scientific community understood the importance, novelty and contribution of fuzzy logic. Therefore, the first published work of Lotfi Zadeh on fuzzy sets [2] had 63,612 citations up to 4 August 2023, according to the Scopus scientific peer-reviewed published studies database. Since then, fuzzy logic has been studied by several researchers and used in control systems and complex systems that have to take a decision and give an output based on vague and inaccurate data. For complex systems, Zadeh stated that "as the complexity of a system increases, our ability to make precise and yet significant statements about its behavior diminishes until a threshold is reached beyond which precision and significance (or relevance) become almost mutually exclusive characteristics" [4]. As a consequence, the term "Fuzzy Inference System" was derived. A Fuzzy Inference System is any system that uses fuzzy logic and rules to reason and/or take a decision. It consists of four modules (Fig 1): (i) fuzzification: it converts input data into fuzzy sets; (ii) fuzzy rules base: a set of IF-THEN rules with linguistic variables and values; (iii) the inference engine: it decides which rules to activate for the input values, and then, execute and combine them to 'calculate' the result; (iv) defuzzification: it converts the fuzzy value of the result into a crisp value.

The term "Fuzzy Inference System" includes Fuzzy Logic-based software systems, which concern any automated program, process or application that uses Fuzzy Logic to perform reasoning and/or decision making. Fuzzy logic-based software systems are used in medicine, education, natural language processing, automotive engineering, business processes, consumer electronics (like air-conditioning, dishwashers and heating), decision making and recommendation, control systems, etc. For example, in an automotive system, fuzzy logic is used for speed and traffic control; in dishwashers, fuzzy logic is used to determine the washing strategy and

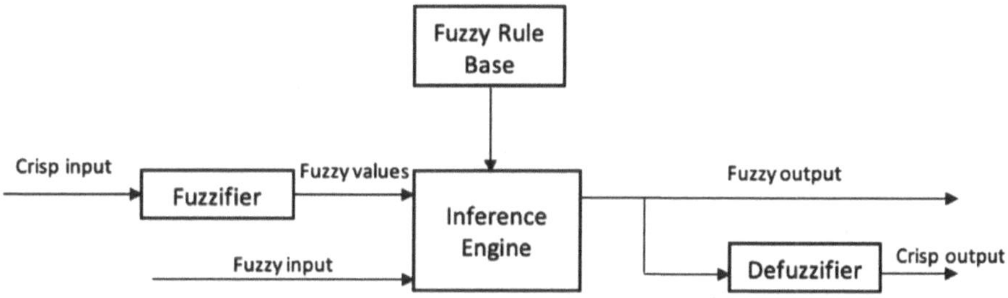

Fig. 1 The architecture of a Fuzzy Inference System

Table 2 Statistics concerning published studies on fuzzy logic

Keyword	Published research document results	Period
Fuzzy logic	102,262	From 1953 up to August 2023
	25,969	Last 5 years-from 2019 to August 2023
Fuzzy inference system	21,141	From 1989 up to August 2023
	7,908	Last 5 years-from 2019 to August 2023
Fuzzy systems	61,693	From 1969 up to August 2023
	13,543	Last 5 years-from 2019 to August 2023

the power needed, according to which the dishes load and the food scraps on them; in medicine, fuzzy logic is used for automatic and computer-aided diagnoses; in natural language processing, fuzzy logic is used to determine semantic relations between words.

In recent years, the increased interest in Artificial Intelligence has led researchers to work actively with fuzzy logic. According to the Scopus scientific peer-reviewed published studies database, there is an increased scientific interest in Fuzzy Logic (Table 2). 26.70% of the total published research manuscripts concerning Fuzzy Logic have been published in the last 5 years. Similarly, 39.43% of the overall published scientific research papers, which concerns Fuzzy Inference Systems, have been published in the last five years. The corresponding percentage for the published research studies on fuzzy systems is 25.21%. The reasons for the increased interest in fuzzy logic and systems are the following.

1. Fuzzy logic manages effectively vague and imprecise data.
2. The mathematics of fuzzy logic is simple.
3. Fuzzy logic and rules are understandable.

4. Fuzzy logic is flexible.
5. Fuzzy logic imitates the human way of thinking and reasoning.
6. Fuzzy logic can be combined with other AI techniques.

The main body of this work is organized into the following five chapters. In Chap. 1, an introduction to fuzzy logic, fuzzy sets, fuzzy operators, membership functions, fuzzy rules and fuzzy inference systems is presented. In Chap. 2, the role of fuzzy logic in artificial intelligence and its use in smart applications are described. In Chap. 3, several fuzzy logic-based software systems in the discipline of medicine, education, decision making and recommendation, natural language processing, automotive engineering and industry, heating, ventilation and air-conditioning, navigation, scheduling, network traffic and security are presented. In Chap. 4, a comparative discussion about the use of fuzzy logic in smart applications and systems of a variety of fields is presented. Finally, in Chap. 5, conclusions are drawn and discussed.

Piraeus, Greece Konstantina Chrysafiadi

References

1. F.J. Pelletier, Review of metamathematics of fuzzy logics (PDF). Bull. Symb. Log. **6**(3), 342–346. https://doi.org/10.2307/421060. JSTOR 421060. Archived (PDF) from the original on 03 Mar 2016 (2000)
2. L.A. Zadeh, Fuzzy Sets. Information and Control, **8**(3), 338–353 (1965)
3. L.A. Zadeh, Quantitative fuzzy semantics. Inf. Sci., **3**(2), 159–176 (1971)
4. L.A. Zadeh, Outline of a new approach to the analysis complex systems and decision processes. IEEE Trans. Syst. Man. Cybernet. **3**, 28–44 (1973)

Contents

Abbreviations

AHRS	Attitude-Heading Reference System
AI	Artificial Intelligence
ANFIS	Adaptive Neuro Fuzzy Inference System
CPU	Central Processing Unit
CT	Computed Tomography
FAHP	Fuzzy Analytic Hierarchy Process
FCE	Fuzzy Comprehensive Evaluation
FCM	Fuzzy Cognitive Map
FIS	Fuzzy Inference System
FSVM	Fuzzy Support Vector
GNSS	Global Navigation Satellite System
GPS	Global Positioning System
HFLTS	Hesitant Fuzzy Linguistic Term Set
HVAC	Heating, Ventilation and Air Conditioning
INS	Inertial Navigation System
IPFCM	Intuitionist Possibilistic Fuzzy Clustering
IT2FLS	Interval Type-II Fuzzy Logic System
LLSM	Logarithmic Least Square Method
MANET	Mobile Adhoc NETwork
MCDM	Multi-Criteria Decision-Making Method
MRA	Magnetic Resonance Angiography
MRI	Magnetic Resonance Image
NLP	Natural Language Processing
QoS	Quality of Service
RFID	Radio Frequency Identification
SINS	Strap-Down Inertial Navigation System
TOPSIS	Technique for Order of Preference by Similarity to Ideal Solution
URL	Uniform Resource Locator

USBL	Ultra-Short Baseline
VANET	Vehicular Ad hoc NETwork
6LoWPAN	IPv6 over Low-power Wireless Personal Area Networks

Chapter 1
Fuzzy Logic

Abstract Fuzzy logic was introduced by Lotfi A. Zadeh in 1965 as a methodology for computing with words and since then it is used extensively. It is a generalization of the Binary (Boolean) logic, which can manage effectively data that deal with uncertainty, vagueness and subjectivity. In this chapter basic concepts of fuzzy logic, fuzzy sets and membership functions are presented. Also, the structure and operation of a Fuzzy Inference System (FIS) and the different types of FIS are described.

1.1 Introduction

Fuzzy logic is a generalization of the Binary (Boolean) logic, which can manage effectively data that deal with uncertainty, vagueness and subjectivity. It was introduced by Lotfi A. Zadeh in 1965 as a methodology for computing with words [1]. According to fuzzy logic an element can belong to a set in a degree between 0 (not at all) and 1 (absolutely). When, according to the Binary (Boolean) logic, an element can belong absolutely to a set or does not belong to the set. There are no intermediate states. This difference is illustrated in Fig. 1.1.

In other words, a logical sentence can be true or false according to the Binary (Boolean) logic. For example, "the dolphin is a mammal" is true, 'the dog flies" is false, "2 + 3 = 5" is true, "3 * 8 = 10" is false. However, according to Fuzzy Logic a logical sentence can be true in a degree of truth. For example, "a man 15 years old is young" is 100% true (truth value = 1), "when the temperature is 26 °C, then it is warm" is 70% true (truth value = 0.7), "when the car's speed is 65 km/h, then it goes fast" is 50% true (truth value = 0.5), "the sky is green" is 0% true" (truth value = 0). Fuzzy logic uses linguistic values to describe data, situations and conditions.

1.1.1 Fuzzy Sets and Basic Concepts

A fuzzy set is a set of objects, each one of which were characterized by a function that associates to each object a real number in the interval [0, 1], which indicates the degree

K. Chrysafiadi, *Fuzzy Logic-Based Software Systems*, Learning and Analytics in Intelligent Systems 34, https://doi.org/10.1007/978-3-031-44457-9_1

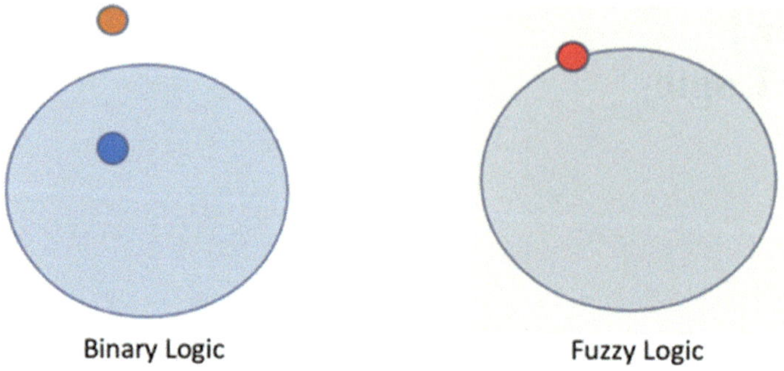

Fig. 1.1 Binary logic versus fuzzy logic

in which the object belongs to the set. This function is called "membership function" and its value is called "membership value" or "membership degree". Therefore, fuzzy set A is called a set of ordered pairs (x, $\mu_A(x)$), where x Î X (a space of objects) and $\mu_A(x)$ Î [0,1]. μ_A is the membership function of A and $\mu_A(x)$ is the membership value of x to A. If $\mu_A(x)$ equals to 0, then it means that x does not belong to A. While, if $\mu_A(x)$ equals to 1, then it means that x belongs fully to A. If $0 < \mu_A(x) < 1$, then it means that x belongs partially to A. Some examples of fuzzy sets are the following:

- Warm = {(5, 0), (10, 0), (15, 0.25), (20, 0.5), (25, 0.75), (30, 1)}
- Young-aged = {(1,0), (5, 0.21), (10, 0.47), (15, 0.74), (20, 1), (25, 1), (30, 0.8), (35, 0.6), (40, 0.4), (45, 0.2), (50, 0)}
- Good_Knowledge_level = {(0,0), (20, 0), (50, 0), (60, 0), (65, 0.25), (70, 0.5), (75, 0.75), (80, 1), (85, 0.5), (90, 0), (100, 0)}
- Fast = {(0, 0), (20, 0), (40, 0), (60, 0.25), (80, 0.5), (100, 0.75), (120, 1), (140, 1), (160, 1), (180, 1)}.

In Fig. 1.2 the graphical representation of the above fuzzy sets is depicted.

A membership function calculates the membership degree of each fuzzy set's object to the fuzzy set. As input, it takes all the possible values of the set's members and as output it gives a value between 0 and 1. In its graphical representation, the horizontal axis (axis X) concerns the possible members of the set and the vertical axis (axis Y) has values from 0 to 1. The most common used types of membership functions are the following:

- Triangular membership function

This is one of the most common used membership functions in fuzzy systems. It is called triangular because in the graphical representation, it designs a triangle. It is defined by three parameters a, b and c. a is the first object (with lower value) that belongs to the fuzzy set, b is the object whose membership value is 1 (the value that corresponds to the top of the designed triangle), and c is the last object (with higher value) that belongs to the fuzzy set (Fig. 1.3). A triangular membership function is

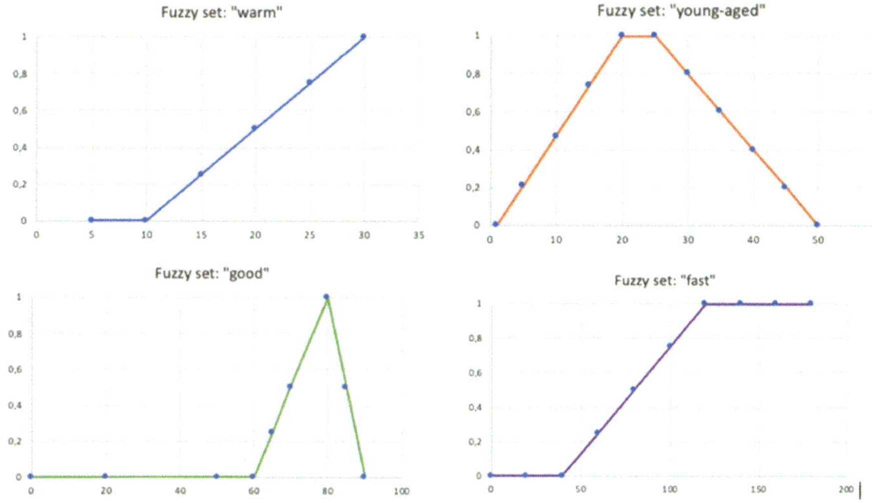

Fig. 1.2 Examples of fuzzy sets

defined as follows:

$$\mu_{\mathrm{A}}(x) = \begin{cases} 0, if\ x \leq a\ or\ x \geq c \\ \frac{x-a}{b-a}, if\ a \leq x \leq b \\ \frac{c-x}{c-b}, if\ b \leq x \leq c \end{cases}$$

An example of triangular fuzzy sets is the phases of a 24 h day. We define the following fuzzy sets: {morning, noon, afternoon, evening, midnight}. The set of input values is [0, 24]. The membership functions of the above fuzzy sets and their graphical representation are depicted in Figs. 1.4 and 1.5.

- Trapezoidal membership function

This is another most common used membership functions in fuzzy systems. It is called trapezoidal because in the graphical representation, it designs a trapezoid. It is defined by four parameters a, b, c and d. a is the first object (with lower value—the

Fig. 1.3 Triangular membership function

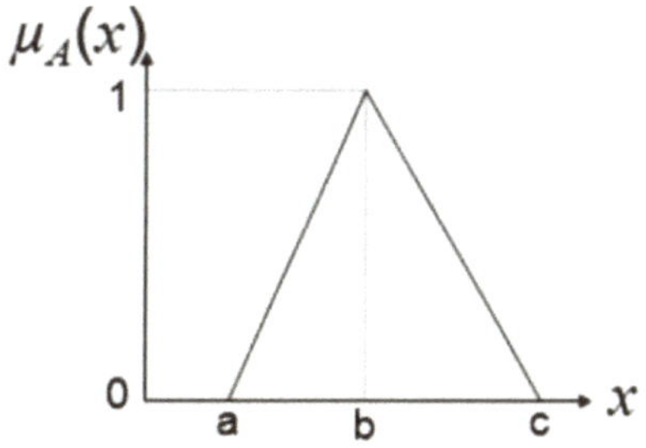

$$\mu_{morning} = \begin{cases} 0, & x = 0 \text{ ή } x \geq 14 \\ \frac{x}{8}, & 0 \leq x \leq 8 \\ \frac{14-x}{6} & ,8 \leq x \leq 14 \end{cases} \qquad \mu_{noon} = \begin{cases} 0, & x \leq 8 \text{ ή } x \geq 18 \\ \frac{x-8}{6}, & 8 \leq x \leq 14 \\ \frac{18-x}{4} & ,14 \leq x \leq 18 \end{cases}$$

$$\mu_{afternoon} = \begin{cases} 0, & x \leq 14 \text{ ή } x \geq 20 \\ \frac{x-14}{4}, & 14 \leq x \leq 18 \\ \frac{20-x}{2} & ,18 \leq x \leq 20 \end{cases} \qquad \mu_{evening} = \begin{cases} 0, & x \leq 18 \text{ ή } x = 24 \\ \frac{x-18}{2}, & 18 \leq x \leq 20 \\ \frac{24-x}{4} & ,20 \leq x \leq 24 \end{cases}$$

$$\mu_{midnight} = \begin{cases} 0, & x = 0 \text{ ή } x = 24 \\ \frac{x-20}{4}, & 20 \leq x \leq 24 \\ \frac{8-x}{8} & ,0 \leq x \leq 8 \end{cases}$$

Fig. 1.4 The triangular membership function of a 24 h

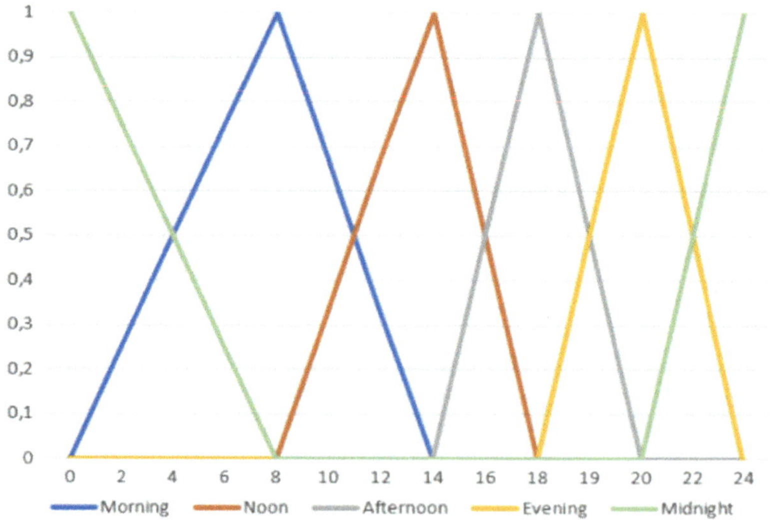

Fig. 1.5 Graphical representation of fuzzy sets that describe a 24 h

start of the base of the designed trapezoid) that belongs to the fuzzy set, b is the first object whose membership value is 1 (the value that corresponds to the height of the designed trapezoid), c is the last object whose membership value is 1 (it belongs absolutely to the membership set), and d is the last object (the end of the base of the designed trapezoid) that belongs to the fuzzy set (Fig. 1.6). A trapezoidal membership function is defined as follows:

$$\mu_{\mathrm{A}}(x) = \begin{cases} 0, if\ x \le a\, or\, x \ge d \\ \frac{x-a}{b-a}, if\ a \le x \le b \\ 1, if\ b \le x \le c \\ \frac{d-x}{d-c}, if\ c \le x \le d \end{cases}$$

An example of trapezoidal fuzzy sets is the human's age. We define the following fuzzy sets: {infant, child, young, middle-aged, senior}. The set of input values is [0, 110]. The membership functions of the above fuzzy sets and their graphical representation are depicted in Figs. 1.7 and 1.8.

Fig. 1.6 Trapezoidal membership function

$$\mu_{infant} = \begin{cases} 0, & x \ge 7 \\ \frac{x}{3}, & x \le 3 \\ \frac{7-x}{2} & ,5 \le x \le 7 \end{cases} \qquad \mu_{child} = \begin{cases} 0, & x \le 5\ ή\ x \ge 18 \\ \frac{x-5}{2}, & 5 \le x \le 7 \\ 1, & 7 \le x \le 12 \\ \frac{18-x}{4} & ,14 \le x \le 18 \end{cases} \qquad \mu_{young} = \begin{cases} 0, & x \le 12\ ή\ x \ge 50 \\ \frac{x-12}{6}, & 12 \le x \le 18 \\ 1, & 18 \le x \le 35 \\ \frac{50-x}{15} & ,35 \le x \le 50 \end{cases}$$

$$\mu_{middle-aged} = \begin{cases} 0, & x \le 35\ ή\ x = 80 \\ \frac{x-35}{15}, & 35 \le x \le 50 \\ 1, & 50 \le x \le 65 \\ \frac{80-x}{15} & ,65 \le x \le 80 \end{cases} \qquad \mu_{senior} = \begin{cases} 0, & x \le 65 \\ \frac{x-65}{15}, & 65 \le x \le 80 \\ 1\ , & x \ge 80 \end{cases}$$

Fig. 1.7 The trapezoidal membership function of a human's age

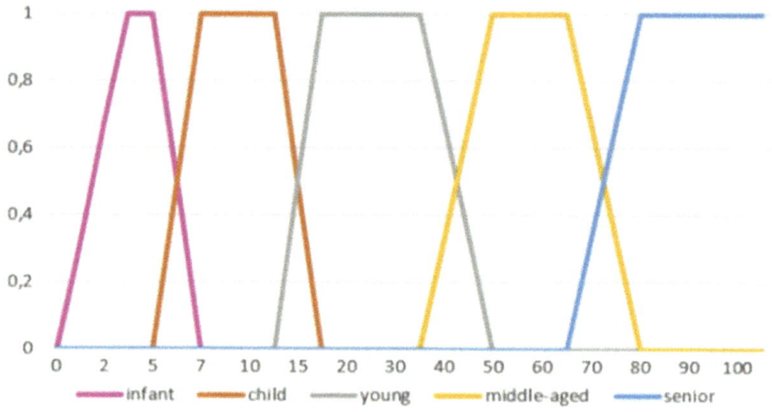

Fig. 1.8 Graphical representation of fuzzy sets that describe a human's age

- Generalized bell-shaped Membership Function

Generalized bell-shaped membership function designs is a symmetrical shape, which resembles to a bell. It is described by three parameters a, b, and c. a determines the designed curve's width. b is a positive integer that indicates the slop of the designed curve. Therefore, the value of b determines the rate of function's decline from 1 to 0. So, as the value of b increases, the rate of function's decline increases too, and the designed curve becomes steeper. In analogy, as the value of b decreases, the rate of function's decline decreases too. Thus, the value of parameter b indicated the degree of fuzziness of the defined fuzzy set. The lower the value of parameter b is, the greater the ambiguity of the fuzzy set is. Similarly, the higher the value of parameter b is, the less the ambiguity of the fuzzy set is. Consequently, when high values are assigned to b, the fuzzy set tends to become crisp. c is the center of the curve. The graphical representation of a generalized bell-shaped membership function is depicted in Fig. 1.9. There are two points of inflection of the designed curve: c − a and c + a. The curve's slope to the infection points is $-\frac{b}{2a}$. A generalized bell-shaped membership function is defined as follows:

$$\mu_{\mathrm{A}}(x) = \frac{1}{1 + \left(\frac{x-c}{a}\right)^{2b}}$$

An example of generic bell-shaped fuzzy sets is the following, which used to describe the temperature. We define the following fuzzy sets: {cold, warm, hot}. The set of input values is [−10, 45]. The membership functions of the above fuzzy sets and their graphical representation are depicted in Figs. 1.10 and 1.11.

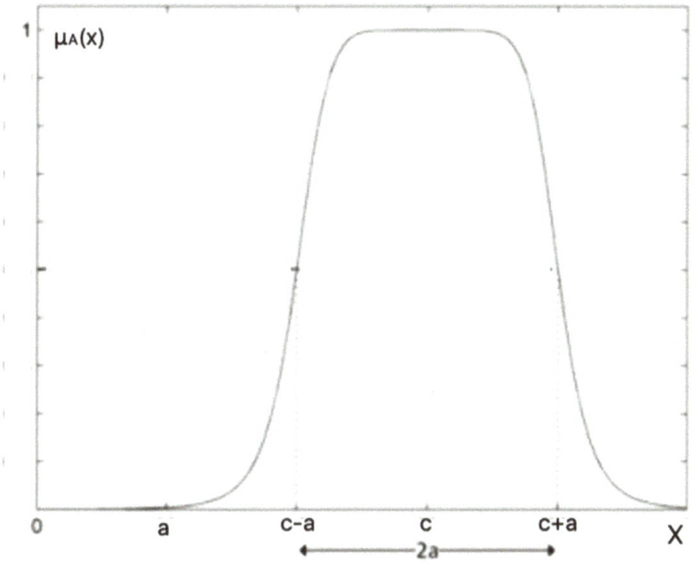

Fig. 1.9 Generalized bell-shaped membership function

$$\mu_{cold}(x) = \frac{1}{1 + (\frac{x+10}{15})^6} \qquad \mu_{warm}(x) = \frac{1}{1 + (\frac{x-25}{10})^6} \qquad \mu_{hot}(x) = \frac{1}{1 + (\frac{x-42}{8})^6}$$

Fig. 1.10 The generalized bell-shaped membership functions of temperature

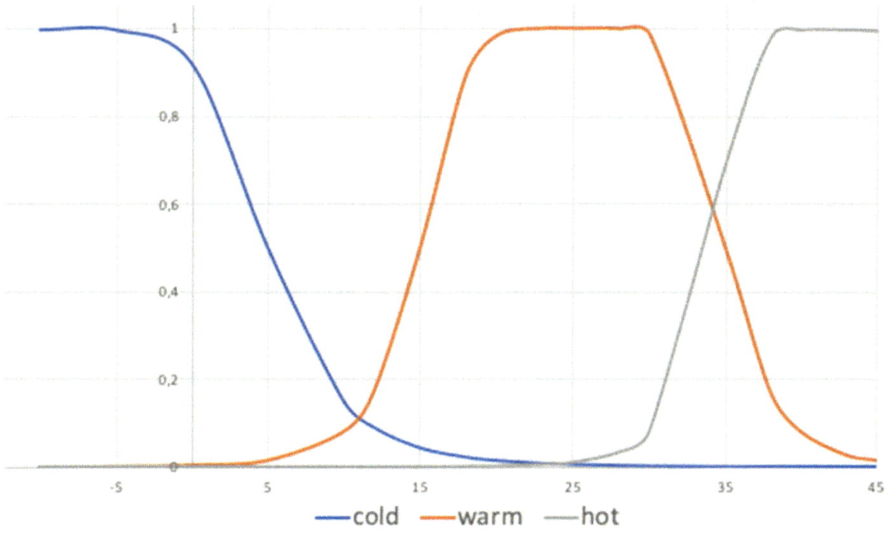

Fig. 1.11 Graphical representation of fuzzy sets that describe temperature

- Gaussian membership function

Gaussian membership function is, also, symmetric. It is described by two parameters a and b. The parameter a is the center of the curve that is designed by the Gaussian membership function. The parameter b is the curve's width. The two points of inflection of the designed curve are a − b and a + b. The graphical representation of a Gaussian membership function is depicted in Fig. 1.12. A Gaussian membership function is defined as follows:

$$\mu_A(x) = e^{-(\frac{x-a}{b})^2}$$

An example of Gaussian fuzzy sets is the description of UV sunlight radiation. We define the following fuzzy sets: {low, medium, high}. The set of input values is [0, 11] and corresponds to the UV indexes. The membership functions of the above fuzzy sets and their graphical representation are depicted in Figs. 1.13 and 1.14.

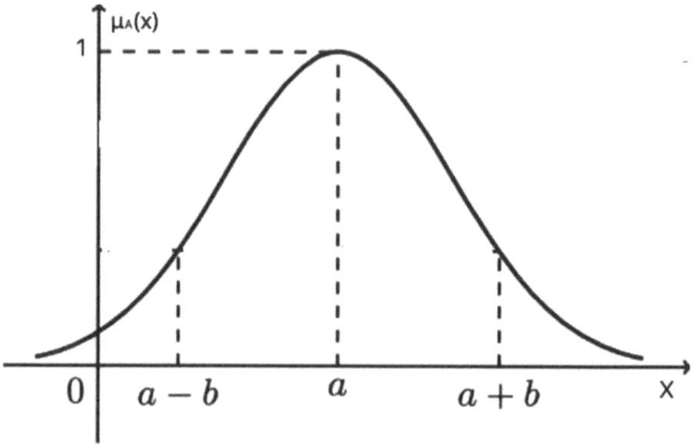

Fig. 1.12 Gaussian membership function

$$\mu_{low}(x) = e^{-(\frac{x}{2})^2} \qquad \mu_{medium}(x) = e^{-(\frac{x-5}{2})^2} \qquad \mu_{high}(x) = e^{-(\frac{x-11}{2})^2}$$

Fig. 1.13 The Gaussian membership functions of UV sunlight radiation

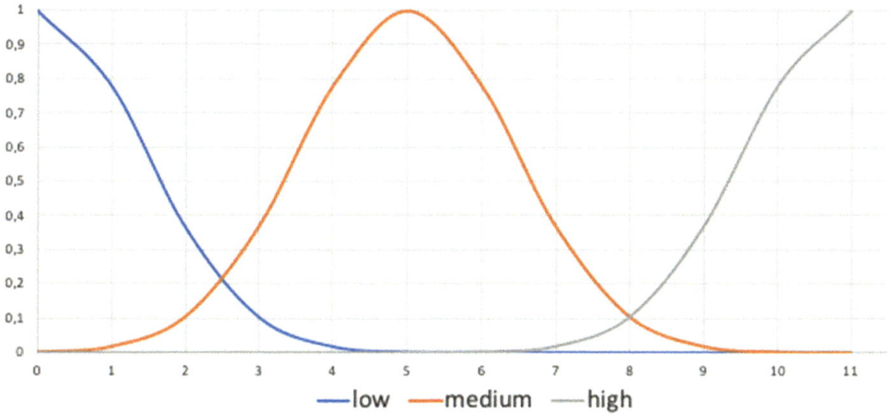

Fig. 1.14 Graphical representation of fuzzy sets that describe UV sunlight radiation

1.1.2 Basic Operations of the Fuzzy Sets

Let's consider the two following fuzzy sets:

- $A = (x, \mu_A(x)), \mu_A(x) \in [0,1]$
- $B = (x, \mu_B(x)), \mu_B(x) \in [0,1]$.

The following operations between them stand.

- Complement Ã του A: $\mu_{\tilde{A}}(x) = 1\text{-}\mu A(x), \forall\, x \in X$ (Fig. 1.15)

- Union A È B: $\mu_{A\grave{E}B}(x) = \max(\mu_A(x), \mu_B(x)), \forall\, x \in X$ (Fig. 1.16)

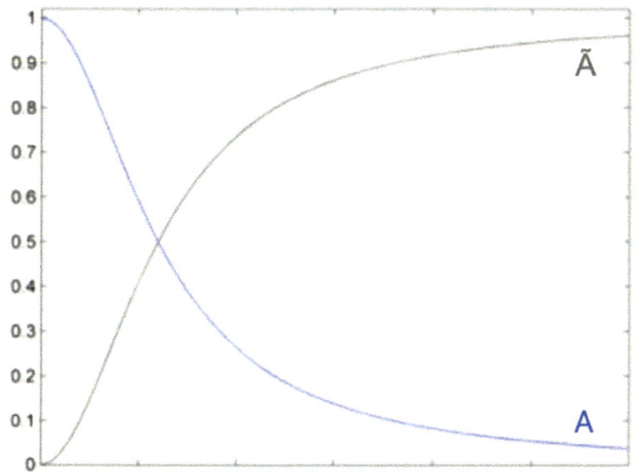

Fig. 1.15 Complement of fuzzy set A

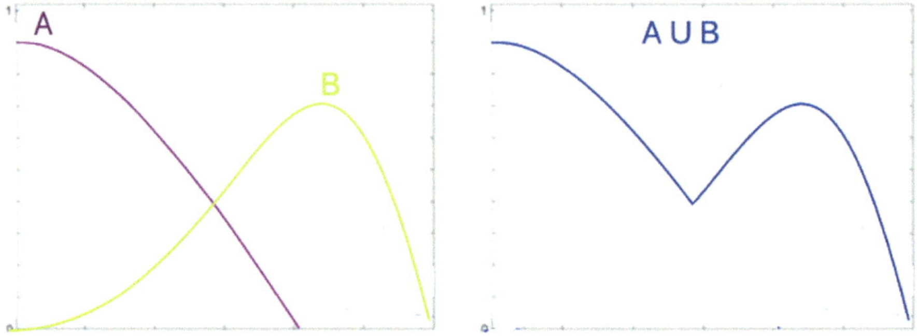

Fig. 1.16 Union of fuzzy sets A and B

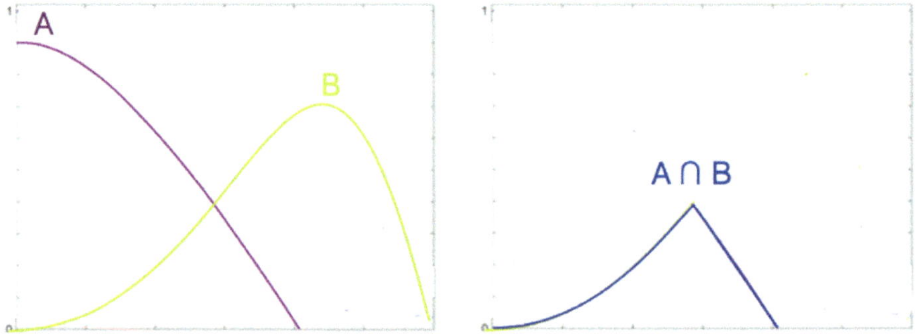

Fig. 1.17 Intersection of fuzzy sets A and B

- Intersection A ∩ B: $\mu_{A \cap B}(x) = \min(\mu_A(x), \mu_B(x))$, $\forall\ x \in X$ (Fig. 1.17)

- Cartesian product A x B: $\mu_{AxB}(\times 1, \times 2) = \min(\mu_A(\times 1), \mu_B(\times 2))$, $\times\ 1 \in X1$ (definition field of A) και $\times\ 2 \in X2$ (definition field of B).

1.1.3 Relationships Between Fuzzy Sets

Let's consider the two following fuzzy sets:

- $A = (x, \mu_A(x)), \mu_A(x) \in [0,1]$
- $B = (x, \mu_B(x)), \mu_B(x) \in [0,1]$.

The following relationships between them stand.

- Equality: A is equal to B ($A = B$), if $\mu_A(x) = \mu_B(x)$, $\forall\ x \in X$.

The similarity measure is used to calculate the closeness between two fuzzy sets. The similarity between two fuzzy sets is calculated by the following formula:

$$E(A, B) = E(B, A) = \frac{|A \cap B|}{|A \cup B|}$$

If A = B, then E(A, B) = E(B, A) = 1.

- Subset: A is subset of B (or B includes A), if $\mu_A(x) \leq \mu_B(x)$, $\forall$ x $\in X$. The degree in which A is included in B is calculated by the formula:

$$S(A, B) = \frac{|A \cap B|}{|A|}$$

If A Í B, then S(A, B) = 1.

1.2 Fuzzy Inference System

A Fuzzy Inference System (FIS) is a system that performs mapping of crisp inputs to outputs using Fuzzy Logic theory and fuzzy logic-based rules. Particularly, a FIS takes crisp data as input, maps them to fuzzy sets, and applies fuzzy rules, which represent the expert's knowledge and reasoning process, over them, and produce fuzzy inferences that transforms them into crisp outputs. The fuzzy rules set is the core of a FIS. A fuzzy rule uses fuzzy variables, linguistic values, hedges and fuzzy sentences. A fuzzy variable takes fuzzy values. Their value field consists of fuzzy sets. Examples of fuzzy variables are the following ones:

- Fuzzy variable: 'age'. Value field: {'baby', 'infant', 'young', 'old'}
- Fuzzy variable: 'temperature'. Value field: {'frosty', 'cold', 'warm', 'hot'}
- Fuzzy variable: 'quality'. Value field: {'bad', 'moderate', 'good', 'excellent'}
- Fuzzy variable: 'knowledge level'. Value field: {'novice', 'moderate', 'good', 'expert'}.

In many circumstances, the value of a fuzzy variable can be of the format 'very good', 'absolutely safe', 'extremely low', 'slightly strong' etc. The words 'very', 'absolutely', 'extremely', 'slightly' either strengthen (e.g. very good) or weaken (e.g. slightly strong) the fuzzy values of the variable. These words are called hedges. For example, the value field for the fuzzy variable 'speed' can be {'very slow', 'slow', 'no slow nor fast', 'slightly fast', 'fast', 'very fast'}. A fuzzy sentence assigns a fuzzy value to a fuzzy variable. For example:

- Fuzzy sentence: "Maria's height is medium". Fuzzy variable: 'height'. Fuzzy value: 'medium'.
- Fuzzy sentence: "John's knowledge performance is excellent". Fuzzy variable: 'knowledge performance'. Fuzzy value: 'excellent'.
- Fuzzy sentence: "Mike's is young". Fuzzy variable: 'age'. Fuzzy value: 'young'.
- Fuzzy sentence: "Tomorrow the weather will be warm". Fuzzy variable: 'temperature'. Fuzzy value: 'warm'.

All the above are used in fuzzy rules of a FIS, which is used to calculate outputs and take decisions like a human by computing with words. A fuzzy rule of a FIS has, usually, the format: "if x is A then y is B", where x and y are fuzzy variables and A and B are fuzzy values, which correspond to fuzzy sets. The main difference between a fuzzy and a classical rule is that the fuzzy rule is described by linguistic values and imitate the human way of thinking. For example:

- Classical rule: "if the student's performance is less than 60/100, then the exercises have to be of level 1".

Fuzzy rule: "if the student's performance is low, then the exercises have to be easy".

- Classical rule: "if the temperature is less than 15 °C, then set the thermostat to 20 °C at least and the fan to 100 turns at least".

Fuzzy rule: "If the temperature is low, then set the thermostat to hot and the fan to fast".

The rules are derived from the experts' knowledge or from datasets of numerical data. The set of rules of a FIS has to fulfil the following properties:

- **Completeness**: for any input of the FIS, at least one rule has to be defined.
- **Consistency**: No conflicting rules (rules that have the same hypothetical part and a different conclusion) have to exist.
- **Continuity**: The set of rules has to be continuous.

A FIS consists of four modules: (i) fuzzification: converts input data into fuzzy sets, (ii) fuzzy rules base: a set of IF–THEN rules with linguistic variables and values, (iii) the inference engine: it decides which rules to activate for the input values, and then, execute and combine them to 'calculate' the result, (iv) defuzzification: it converts the fuzzy value of the result into a crisp value. The architecture of a FIS is depicted in Fig. 1.18.

There are two categories of FIS: (i) the Mamdani Fuzzy Inference System and (ii) the Takagi–Sugeno Fuzzy Inference System. The main difference between these two categories of FIS is the consequents of their fuzzy rules and the defuzzification process of the fuzzy output. In more details, in Mandani FIS fuzzy rules have the

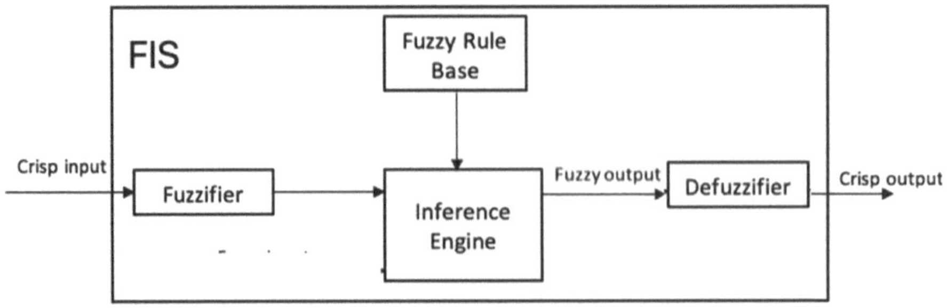

Fig. 1.18 The modules of a Fuzzy Inference System

format: "if input_fuzzy_variable_1 is fuzzy_set_A then output_fuzzy_variable_2 is fuzzy_set_B". The membership values of input_fuzzy_variable_1 to fuzzy_set_A is calculated through membership_function_A. Also, there is membership_function_B to calculate the membership value of fuzzy_variable_2 to fuzzy_set_B". After the execution of the rules that fulfil the input values, the FIS combines the corresponding consequents of the activated rules, produces the fuzzy output and defuzzifies it by applying a defuzzification method (they described below, in the next subsection). On the other hand, in Takagi–Sugeno FIS fuzzy rules have the format: "if input_fuzzy_variable_1 is fuzzy_set_A and input_fuzzy_variable_2 is fuzzy_set_B then output = f(input_fuzzy_variable_1, input_fuzzy_variable_2)". Takagi–Sugeno FIS uses memberships functions to calculate the fuzzy sets of input data, but it does not use membership function in the process of output calculation. The consequent of the rules includes a polynomial function (constant or linear), which involves the defuzzification process. The result of this function is the crisp output of FIS. Below, the two types of FIS are described.

1.2.1 Mamdani Fuzzy Inference System

Mamdani FIS is the most common-used type of FIS. It was proposed by Ebhasim Mamdani in 1975. Initially, it was designed to control the operation of a steam engine and a boiler. A fuzzy rule of a Mamdani FIS has the format: "if x is A then y is B", where x and y are fuzzy variables and A and B are fuzzy values, which correspond to fuzzy sets. In many cases, the operators AND, OR are used to combine inputs and make the fuzzy rules more complex. For example:

- If speed is slow **and** the light distance in short **and** the light color is red, then brake.
- If the project is complex **and** the budget is low, then the risk is very high.
- If the temperature is cold **and** the humidity is normal **or** wet, then the fan's speed has to be slow.

The steps that Mamdani FIS follows to produce the output are the following:

- Step 1: Transform crisp input into fuzzy values calculating the corresponding membership functions.
- Step 2: Check which fuzzy rules are activated for the given values of the input.
- Step 3: Execute the activated rules and extract the consequents.
- Step 4: Combine all the consequents to produce the fuzzy output.
- Step 5: Defuzzify the fuzzy output.

In the next subsections the processes of rules execution and combination and the possible defuzzification methods are presented.

1.2.1.1 Rules Execution and Analysis

Let's have the following fuzzy variables:

- Fuzzy variable x. Fuzzy values: {X1, X2, X3}
- Fuzzy variable z. Fuzzy values: {Z1, Z2, Z3}.

and the fuzzy rule "if x is X1 then z is Z1". Also, we suppose that the membership value of x in X1 is 0.7 ($\mu_{X1}(x) = 0.7$). Then, the output of the rule is that z belongs to Z1 fuzzy set with 0.7 membership degree ($\mu_{Z1}(x) = 0.7$) (Fig. 1.19).

In cases that the operator AND exists in the if part of a fuzzy rule, then either the operator 'min' or the operator 'product' is applied. According to the 'min' operator, the minimum membership value is considered to calculate the membership value of the fuzzy variable in consequent. According to the 'product' operator, the membership value of the fuzzy variable in consequent equals to the arithmetic product of the membership values of the fuzzy variables in in antecedents. For example, let's have the following fuzzy variables:

- Fuzzy variable x. Fuzzy values: {X1, X2, X3}
- Fuzzy variable y. Fuzzy values: {Y1, Y2, Y3}
- Fuzzy variable z. Fuzzy values: {Z1, Z2, Z3}.

and the fuzzy rule "if x is X2 and y is Y3 then z is Z2". Also, we suppose that the membership value of x in X2 is 0.3 ($\mu_{X2}(x) = 0.3$), and the membership value of y in Y3 is 0.8 ($\mu_{Y3}(y) = 0.8$). Then, the output of the rule is that z belongs to Z2 fuzzy set with:

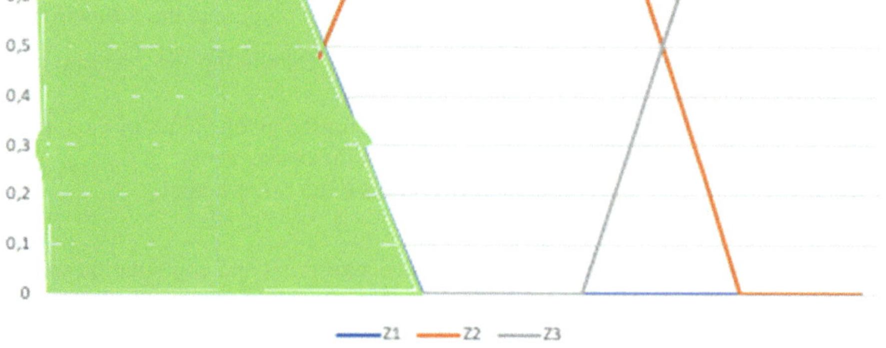

Fig. 1.19 Output of the rule "if x is X1 then z is Z1" with $\mu_{X1}(x) = 0.7$

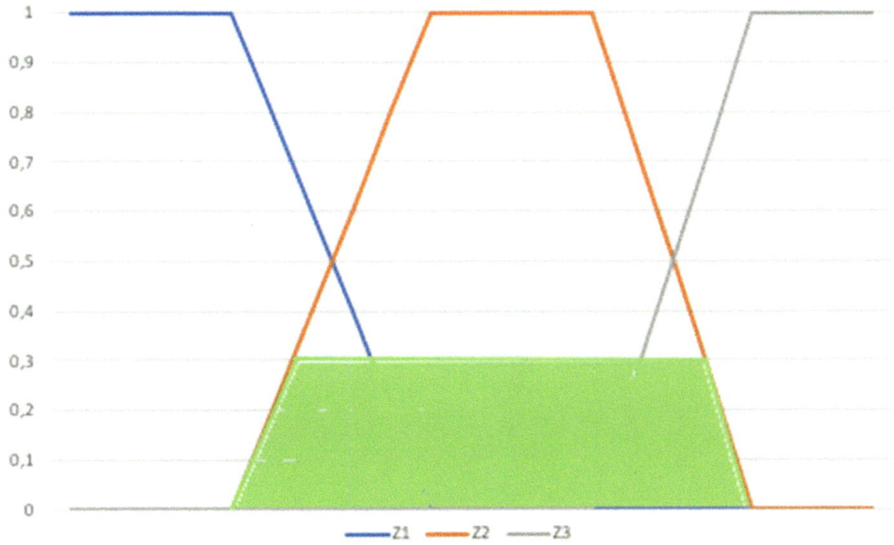

Fig. 1.20 Operator min—output of the rule "if x is X2 and y is Y3 then z is Z2" with $\mu_{X2}(x) = 0.3$ and $\mu_{Y3}(y) = 0.8$

- $\min(\mu_{X2}(x), \mu_{Y3}(y)) = 0.3$ membership value, according to 'min' operator (Fig. 1.20),
- or $\text{product}(\mu_{X2}(x), \mu_{Y3}(y)) = \mu_{X2}(x) * \mu_{Y3}(y) = 0.24$, according to 'product' operator (Fig. 1.21).

In cases that the operator OR exists in the if part of a fuzzy rule, then either the operator 'max' or the operator 'asum' is applied. According to the 'max' operator, the maximum membership value is considered to calculate the membership value of the fuzzy variable in consequent. According to the 'asum' operator, the membership value of the fuzzy variable in consequent equals to the difference of the product of the membership values of the fuzzy variables in in antecedents from their sum. For example, let's have the following fuzzy variables:

- Fuzzy variable x. Fuzzy values: {X1, X2, X3}
- Fuzzy variable y. Fuzzy values: {Y1, Y2, Y3}
- Fuzzy variable z. Fuzzy values: {Z1, Z2, Z3}.

and the fuzzy rule "if x is X1 or y is Y2 then z is Z3". Also, we suppose that the membership value of x in X1 is 0.3 ($\mu_{X1}(x) = 0.3$), and the membership value of y in Y2 is 0.5 ($\mu_{Y2}(y) = 0.5$). Then, the output of the rule is that z belongs to Z3 fuzzy set with:

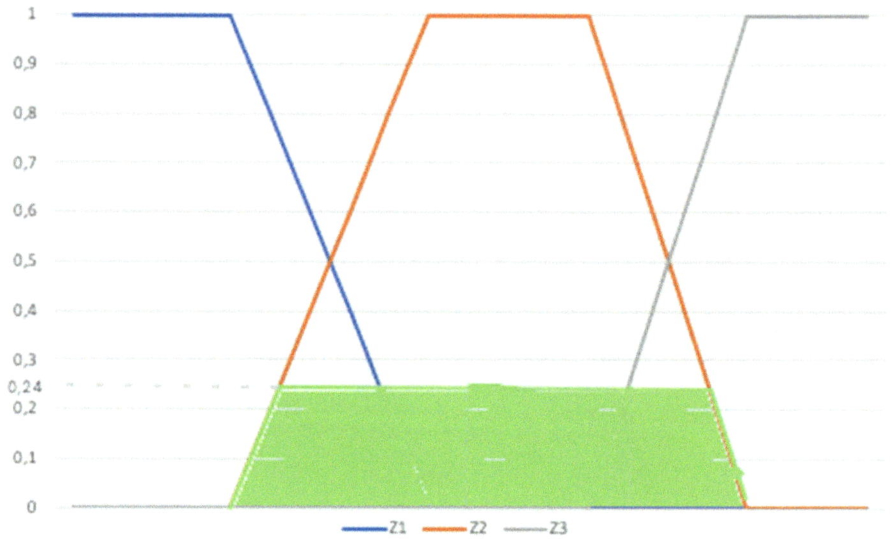

Fig. 1.21 Operator product—output of the rule "if x is X2 and y is Y3 then z is Z2" with $\mu_{X2}(x) = 0.3$ and $\mu_{Y3}(y) = 0.8$

- $\max(\mu_{X1}(x), \mu_{Y2}(y)) = 0.5$ membership value, according to 'max' operator (Fig. 1.22),

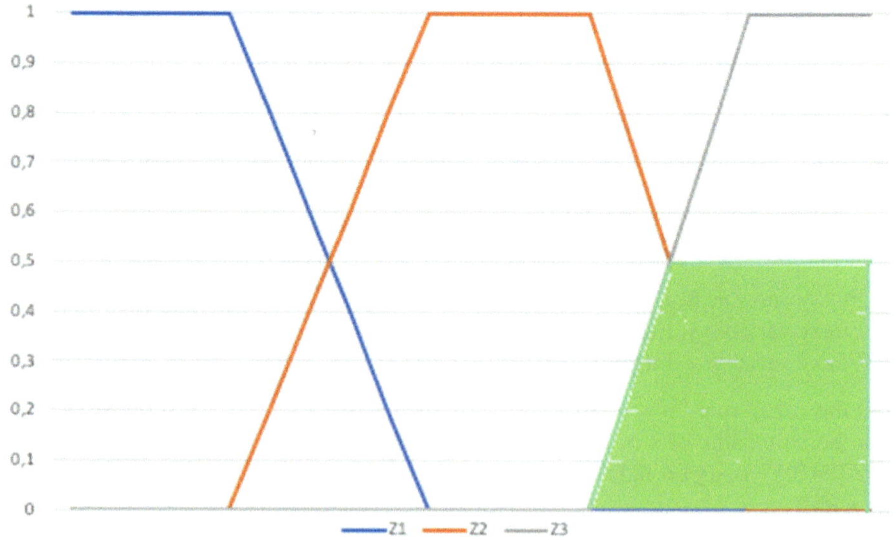

Fig. 1.22 Operator max—output of the rule "if x is X1 or y is Y2 then z is Z3" with $\mu_{X1}(x) = 0.3$ and $\mu_{Y2}(y) = 0.5$

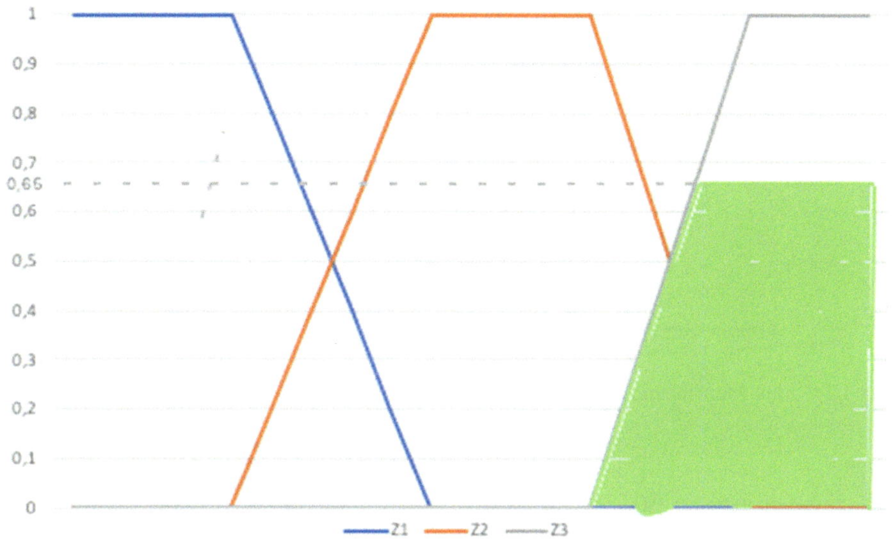

Fig. 1.23 Operator asum—output of the rule "if x is X1 or y is Y2 then z is Z3" with $\mu X1(x) = 0.3$ and $\mu_{Y2}(y) = 0.5$

- or $asum(\mu_{X1}(x), \mu_{Y2}(y)) = (\mu_{X1}(x) + \mu_{Y2}(y)) - (\mu_{X1}(x) * \mu_{Y2}(y)) = 0.65$, according to 'asum' operator (Fig. 1.23).

1.2.1.2 Defuzzification Methods

In the last step of a FIS Mamdani, the fuzzy output has to be defuzzied to give a crisp output. The defuzzification process is based on the membership function of the output fuzzy variable and can be performed by several methods. Below, each one of the defuzzification methods are presented.

(a) Max-Membership Principle

This method is, also, known as Height method. The crisp result is the value of the output variable, for which the membership function has the highest value (Fig. 1.24). It is applicable, mainly, for triangular membership functions. Examples of defuzzification results with max–min principle are depicted in Fig. 1.25.

(b) Centroid Method

This method is the most popular defuzzification method. It finds the center of gravity of the surface between the curve, which is designed by the output fuzzy sets, and

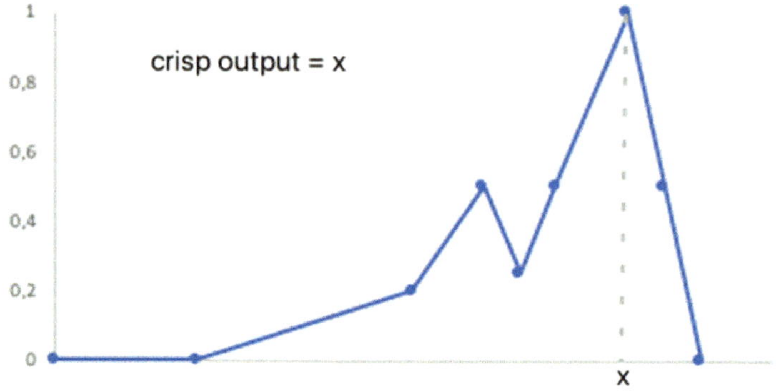

Fig. 1.24 Max-Membership Principle defuzzification method

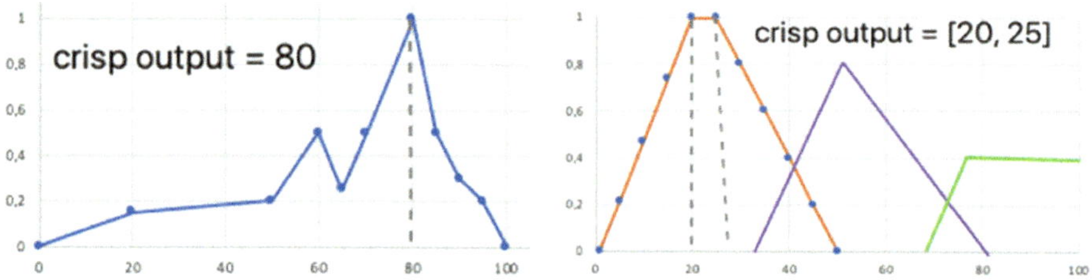

Fig. 1.25 Examples of Max-Membership Principle defuzzification method

axes. The center of gravity is calculated by the typos: $z = \frac{\int \mu_A(x)xdx}{\int \mu_A dx}$. An example of defuzzification result with Centroid method are depicted in Fig. 1.26.

(c) Weighted Average Method

This method can be applied only for membership functions with symmetric result. According to this method, the crisp output is calculated according to the following formula: $z = \frac{\sum \mu_A(\overline{x}) * \overline{x}}{\sum \mu_A(\overline{x})}$, where $\overline{x}$ is the mean value of the fuzzy variable (axis x) and $\mu_A(\overline{x})$ is its membership value in fuzzy set A. An example of defuzzification result with Weighted Average method are depicted in Fig. 1.27.

(d) Mean-Max Membership

This method is, also, called middle of the maxima. It is similar to Max-Membership Principle. The difference is that the value of the fuzzy variable for which the membership value is the maximum, is not unique (Fig. 1.28). An example of defuzzification result with Mean-Max Membership method are depicted in Fig. 1.29.

(e) Centre of Sums

This method uses the algebraic sum of the individual fuzzy subsets instead of the sum of their union. The crisp output is given by the formula $z = \frac{\sum E_A * \overline{x}}{\sum E_A}$, where E_A

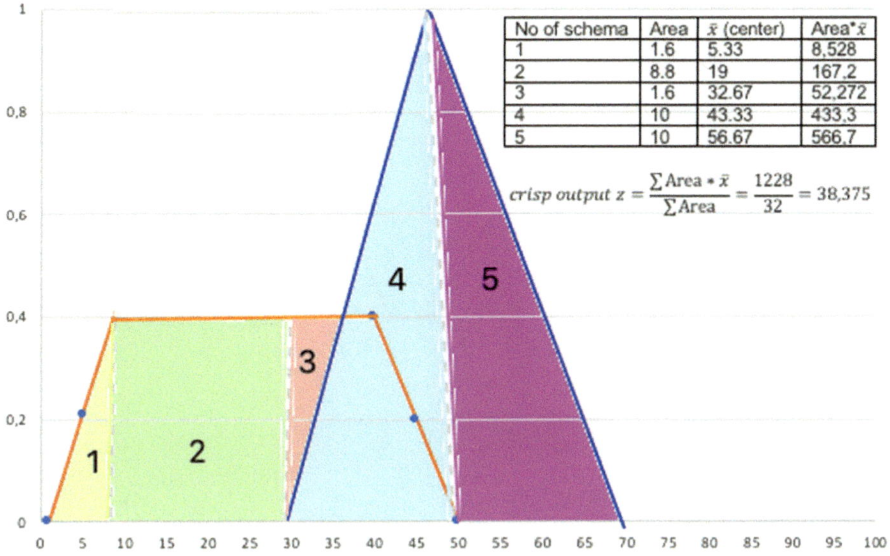

Fig. 1.26 Example of centroid defuzzification method

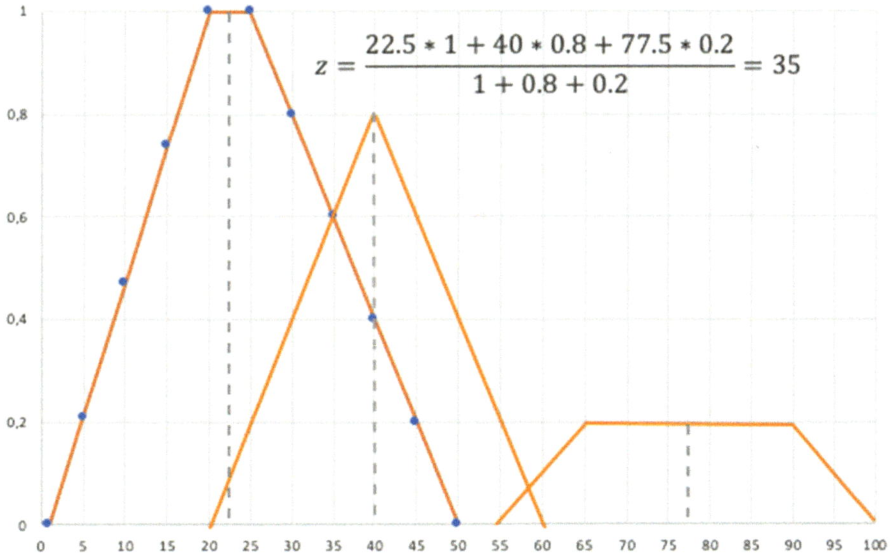

Fig. 1.27 Example of weighted average defuzzification method

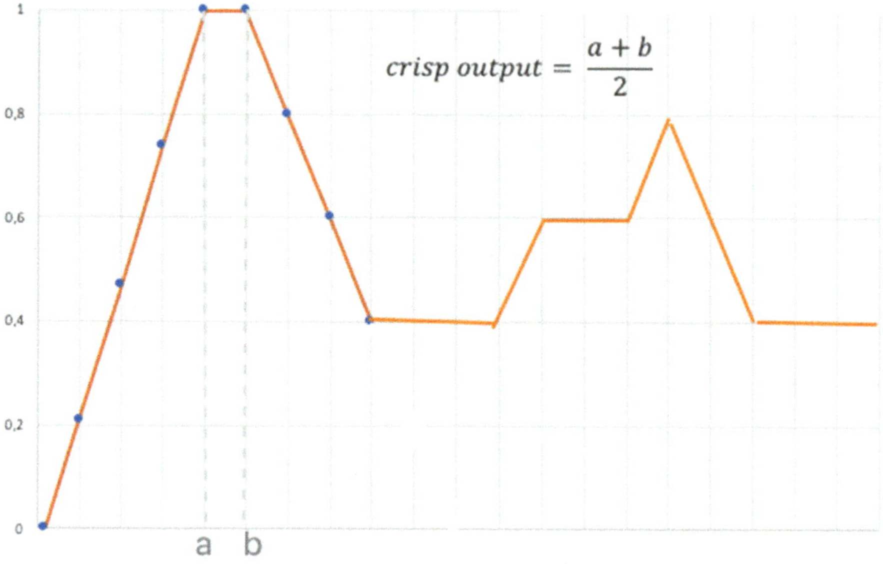

Fig. 1.28 Mean-Max defuzzification method

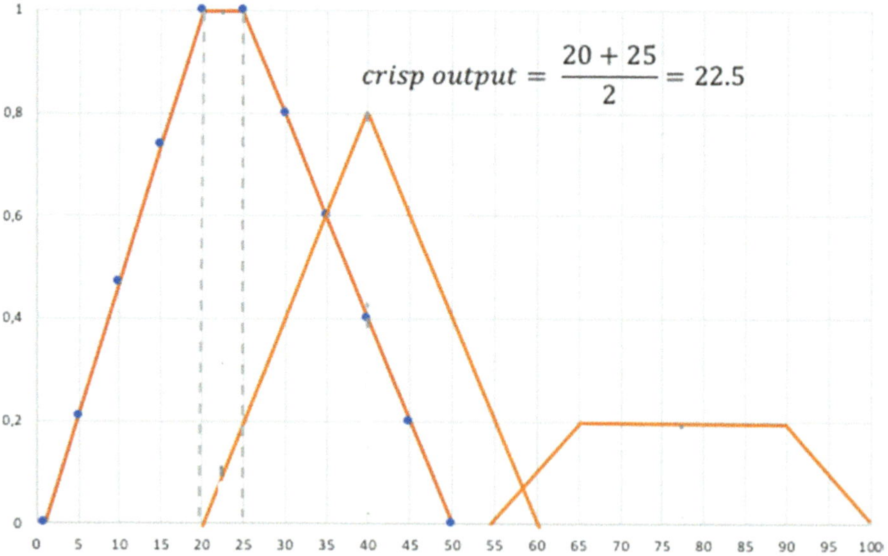

Fig. 1.29 Example of Mean-Max defuzzification method

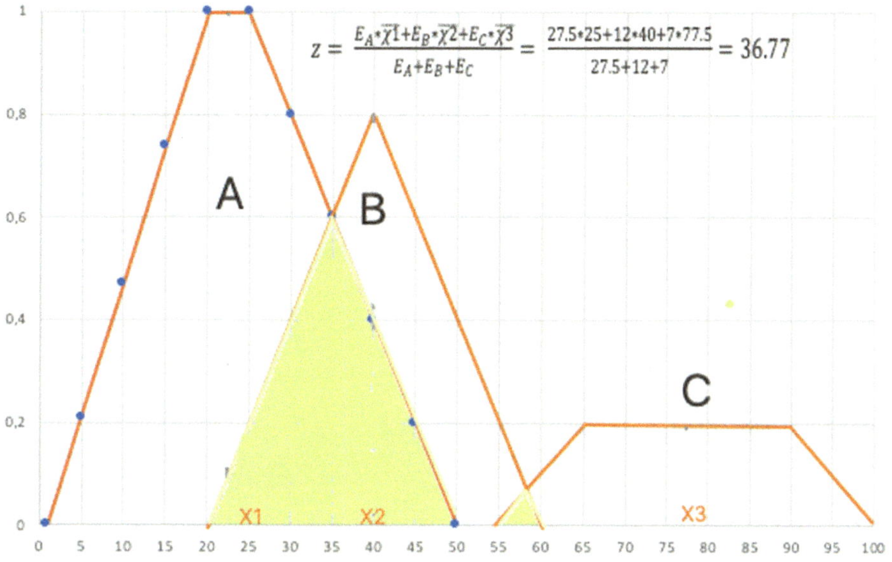

Fig. 1.30 Example of centre of sums defuzzification method

is the area of the designed scheme of the fuzzy set A and $\overline{x}$ is the mean value of the fuzzy variable (axis x) for fuzzy set A. The drawback of this method is that the "common" areas (i.e. the areas, which intersect) are calculated twice. An example of defuzzification result with Centre of Sums method are depicted in Fig. 1.30.

(f) Centre of Largest Area

This method is used in cases, in which there are at least 2 overlapping fuzzy sets. According to this method, the crisp output is the center of gravity of the subset with the maximum area (Fig. 1.31). An example of defuzzification result with Centre of Largest Area method are depicted in Fig. 1.32.

1.2.2 Takagi–Sugeno Fuzzy Inference System

Takagi–Sugeno FIS was introduced by Takagi, Sugeno and Kang in 1985. A fuzzy rule of a Takagi–Sugeno FIS has the format: "if x is A and y is B then z = f(x, y)", where x and y are fuzzy variables and A and B are fuzzy values, which correspond to fuzzy sets. Also, z is the crisp result of the rule, which is calculated by a polynomial function f. Therefore, the defuzzification process is performed in in the consequent of the fuzzy rule. Some examples of Takagi–Sugeno FIS fuzzy rules are the following ones:

- If light distance in short **and** the light color is red, then speed = 40.

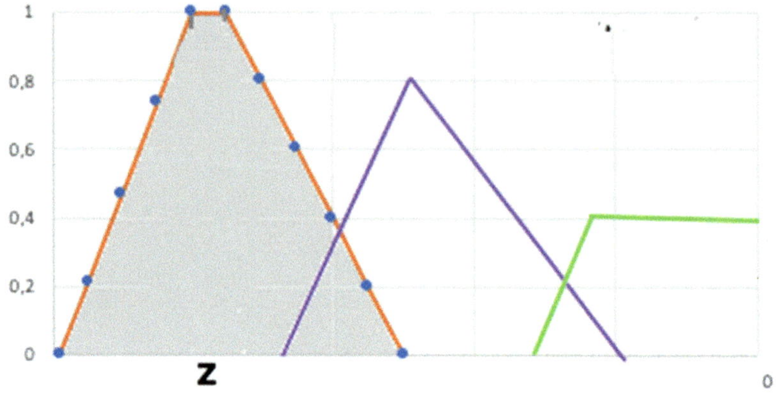

Fig. 1.31 Centre of largest area defuzzification method

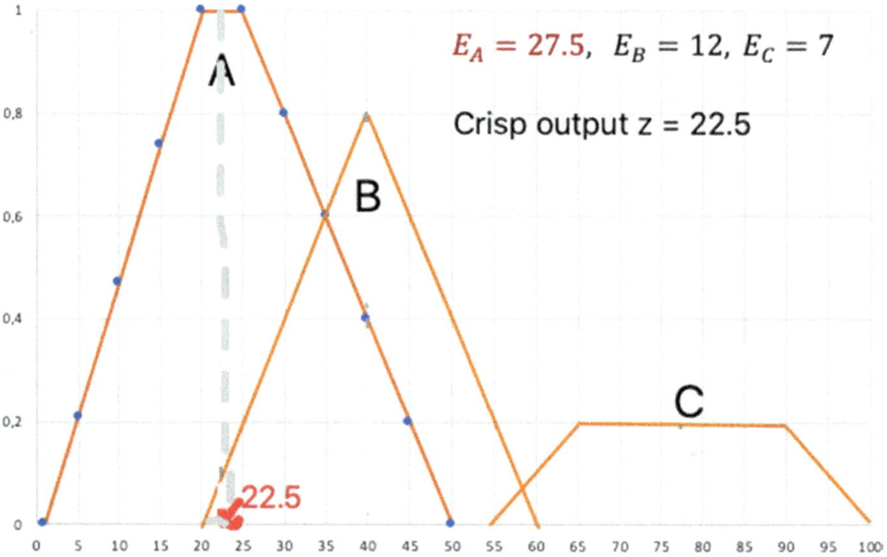

Fig. 1.32 Example of centre of largest area defuzzification method

- If the temperature is very cold **and** the humidity is high, then z = 2 * temperature + humidity.

  The steps that Takagi–Sugeno FIS follows to produce the output are the following:

- Step 1: Transform crisp input into fuzzy values calculating the corresponding membership functions.
- Step 2: Check which fuzzy rules are activated for the given values of the input.
- Step 3: Execute the activated rules and produce the crisp output.

The function in the consequents of fuzzy rules makes Takagi–Sugeno FIS complicated.

1.2.3 Differences Between Mandani FIS and Takagi–Sugeno FIS

Criterion	Mandani FIS	Takagi–Sugeno FIS
Output membership function	Exist	No exist
Defuzzification	It is performed by several defuzzification methods to the fuzzy output	It is included in the function of the consequent of rules
Rules aggregation	Through the selected defuzzification method	Weighted average of the rules' consequent functions result
Modelling	Human way of thinking	Mathematically analysis

Reference

1. L.A. Zadeh, Fuzzy sets. Inf. Control **8**(1965), 338–353 (1965)

Chapter 2
The Role of Fuzzy Logic in Artificial Intelligence and Smart Applications

Abstract Artificial Intelligence (AI) is the process of making machines to think and to do tasks like humans. It includes advanced technologies that make machines more intelligent and capable of performing tasks more efficiently and reliable. There is a variety of AI techniques including machine learning, artificial neural networks, deep learning, knowledge representation and reasoning techniques. An AI technique is Fuzzy Logic, which allows reasoning in cases where there is uncertainty, vagueness, imprecision and/or subjectivity. Fuzzy logic is used in smart applications and systems to make them more intelligent and allow them to reason and behave in a human-like way. Furthermore, in many circumstances fuzzy logic is combined with other AI techniques, like clustering methods, deep learning or artificial neural networks, to create more robust, reliable, effective and adaptable systems. Regarding the above, in this chapter, several fields of fuzzy logic-based software applications and the role of fuzzy logic in them are presented. Also, fuzzy c-means clustering algorithm and adaptive neuro fuzzy inference system technique are described.

Artificial Intelligence (AI) is the process of making machines to think and to do tasks like humans. It includes advanced technologies that make machines more intelligent and capable of performing tasks more efficiently and reliable. There is a variety of AI techniques including machine learning, artificial neural networks, deep learning, knowledge representation and reasoning techniques.

Fuzzy logic is an AI technique, which allows reasoning in cases where there is uncertainty, vagueness, imprecision and/or subjectivity. Fuzzy logic allows handling data and modeling knowledge with linguistic terms, which are called fuzzy sets. Also, it assigns several degrees of truth, which are values between 0 and 1, to each linguistic variable applying corresponding membership functions. This makes fuzzy logic a powerful tool for performing complex processes and making decisions in several real-world applications and problems, which handles uncertain, vague and/or subjective data. This is the reason why fuzzy logic is integrated in software applications and systems of several fields.

Fuzzy logic is used in smart applications and systems to make them more intelligent and allow them to reason and behave in a human-like way. Below, several fields, in which fuzzy logic is used, are presented:

K. Chrysafiadi, *Fuzzy Logic-Based Software Systems*, Learning and Analytics in Intelligent Systems 34, https://doi.org/10.1007/978-3-031-44457-9_2

- Medical systems: Fuzzy logic enables efficient and personalized disease diagnosis and medical image analysis and/or segmentation handling the uncertain and imprecise patient data and images noise.
- Tutoring systems: Fuzzy logic enables a more efficient knowledge and student's characteristics representation, handling the uncertain and imprecise student data. Also, it allows tutoring systems to provide recommendations and personalized solution to students, considering her/his progress, the learning material, the assessment tests etc., mimic the human thinking way. Furthermore, fuzzy logic enables a more efficient and comprehensive student's performance evaluation.
- Decision making support and Recommended systems: Fuzzy logic allows to find the optimal solution in cases where there are uncertain, imprecise, or incomplete data.
- Natural language processing systems: Fuzzy logic enables better understanding and recognize ambiguous or vague words, phrases or language constructs.
- Automotive Systems: Fuzzy logic is used to model driver behavior and/or in several control systems of automotive systems that are used to assist driving, to manage energy efficiency, or to provide ride comfort.
- Scheduling, Network Traffic and Security software systems: Fuzzy logic is used to manage more efficiently the task scheduling processes reducing energy consumption and improving system's performance. Furthermore, fuzzy logic is used for effectively managing the network traffic, optimizing quality of service, and detecting threats and anomalies.
- Navigation systems: Fuzzy logic is used to enhance the complex process of tracking and improve the accuracy in navigation.
- Heating, Ventilation and Air Conditioning (HVAC) systems: Fuzzy logic is used in control systems of HVAC for automatic and adaptive adjustment of temperature, humidity, and air flow and quality, in order to enhance the occupant's thermal comfort. Also, fuzzy logic is used to make smart decisions about energy management and consumption.

Furthermore, in many circumstances fuzzy logic is combined with other AI techniques, like clustering methods, deep learning or artificial neural networks. The aim of such combinations is to create more robust, reliable, effective and adaptable systems. Below, two widely used techniques that combines fuzzy logic with other AI techniques are presented. The first technique is the fuzzy C-means algorithm, which is the combination of the clustering k-means algorithm, with fuzzy logic. The second technique is ANFIS (Adaptive Neuro Fuzzy Inference System), which combines artificial neural network with fuzzy logic.

2.1 Fuzzy C-means

Fuzzy C-means was introduced by by Dunn in 1973 and improved by Bezdek in 1981. The main idea of fuzzy-C means is that each "data item" can belong concurrently to more that one cluster, but in different degrees. The steps of fuzzy c-means clustering are the following:

- **Step 1**: Select a number of clusters and define their centers.
- **Step 2**: Randomly assign coefficients to each "data item" to distribute them to the clusters.
- **Step 3**: Repeat the following process until there is no change in clusters between the last two iterations of the process.
- Calculate the center (centroid) for each cluster.
- Calculate for each "data item" the distance from the center of the cluster.

The fuzzy-C means algorithm is the following:

1. We have a collection of n "data items" $X = \{x_1, x_2, \ldots, x_n\}$
2. We have a collection of c centers of clusters. $C = \{c_1, c_2, \ldots, c_n\}$
3. We create a partition matrix $W = w_{ij} \epsilon [0, 1], i = 1, 2, \ldots., nandj = 1, 2, \ldots., c$

   where w_{ij} is the degree in which the "data item" x_i belongs to the cluster c_j.
4. Calculate the centroid for each cluster according to the formula: $c_k = \frac{\sum_x w_k(x)^m x}{\sum_x w_k(x)^m}$
5. Calculate the distance of each "data item" from the center of each cluster according to the formula: $d_{ij} = \sqrt{\sum_j (x_i - c_{ij})^2}$
6. Recalculate the membership degree values for each "data item" according to the formula $\mu = \sum_1^n \left(\frac{d_{ki}^2}{d_{kj}^2} \right)^{\frac{1}{m} - 1}]^{-1}$
7. Repeat the steps 4 to 6 until there is no change in clusters between the last two iterations of the process.

2.2 Adaptive Neuro Fuzzy Inference System

Adaptive Neuro Fuzzy Inference System (ANFIS) was introduced in the early 1990s. It combines artificial neural networks with fuzzy inference system, and more particular with Takagi–Sugeno fuzzy inference system. ANFIS is an artificial neural network with five levels (Fig. 2.1). ANFIS in Fig. 2.1 has two input variables x_1 and x_2 and one output f. Also, it corresponds to the following fuzzy rules:

- If x_1 is A and x_2 is C, then $f = 0.1 * x_1 + 0.3 * x_2 + 10$
- If x_1 is B and x_2 is D, then $f = x_1 + 0.1 * x_2 + 5$.

The operation of each level is described below.

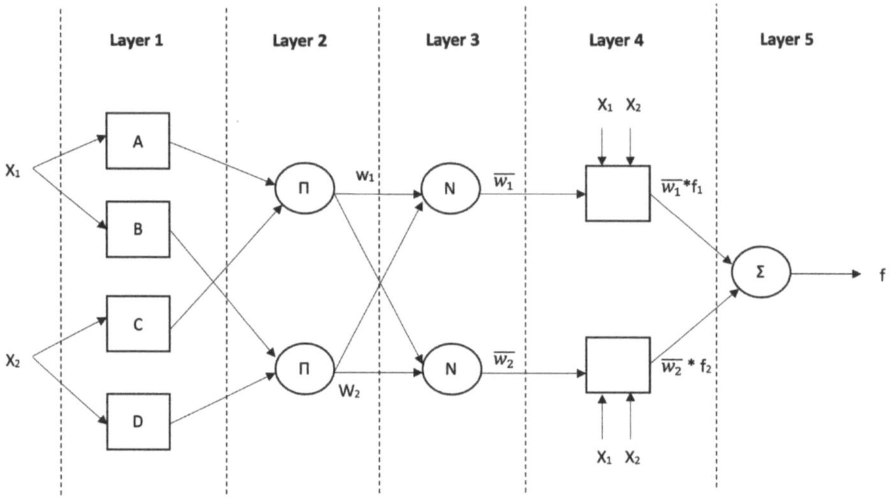

Fig. 2.1 The five levels of ANFIS

Level 1–The Fuzzification Layer

In this level the fuzzification process of the input variables is performed, applying the corresponding membership functions. For example, for the ANFIS in Fig. 2.1, the input variables are x_1 and x_2 and A, B, C, D are the defines fuzzy sets, which correspond to the linguistic values of the input variables. Let's, we have the following membership functions:

$$\mu_A = \begin{cases} 0, x \geq 60 \\ \frac{x}{40}, 0 \leq x \leq 40 \\ \frac{60-x}{20}, 40 \leq x \leq 60 \end{cases} \quad \mu_B = \begin{cases} 0, x \leq 40 or x \geq 100 \\ \frac{x-40}{20}, 40 \leq x \leq 60 \\ \frac{100-x}{40}, 60 \leq x \leq 100 \end{cases}$$

$$\mu_C = \begin{cases} 1, x \leq 50 \\ 0, x \geq 80 \\ \frac{80-x}{30}, 50 \leq x \leq 80 \end{cases} \quad \mu_D = \begin{cases} 0, x \leq 50 \\ 1, x \geq 80 \\ \frac{x-50}{30}, 50 \leq x \leq 80 \end{cases}$$

and $x_1 = 45$ and $x_2 = 70$. Therefore, the output of this level is $\mu_{x1}(A) = 0.75$, $\mu_{x1}(B) = 0.25$, $\mu_{x2}(C) = 0.33$, $\mu_{x2}(D) = 0.67$.

Level 2–The Product Layer

In this level, each node is a fixed node, which is labeled as 'Π'. The output of each node is the product of all the input values. The output value declares the firing strength. For the ANFIS in Fig. 2.1 the following stands:

- $w_1 = \mu_{x1}(A)^* \mu_{x1}(B) = 0.75^*0.25 = 0.1875$
- $w_2 = \mu_{x2}(C)^* \mu_{x2}(D) = 0.33^*0.67 = 0.2211$.

Level 3–The Normalization Layer

In this level, each node is a fixed node, which is labeled as 'N'. The output of each node is the firing strength of the corresponding node 'Π' of the level 2, divided by the sum of all the output values of the previous nodes (firing strengths). The output values of the nodes of level 3 are called normalized firing strengths. For the ANFIS in Fig. 2.1 the following stands:

$$\overline{w_1} = \frac{w_1}{w_1+w_2} = \frac{0.1875}{0.1875+0.2211} = \frac{0.1875}{0.4086} = 0.4589$$
$$\overline{w_2} = \frac{w_2}{w_1+w_2} = \frac{0.2211}{0.1875+0.2211} = \frac{0.2211}{0.4086} = 0.5411.$$

Level 4–The Defuzzification Layer

In this level, each node is an adaptive node, which include a function. This function corresponds to the results-conclusions (the then part) of the fuzzy rules. The output of each node of this level is the product of the function's result with the corresponding normalized firing strength, which was calculated in the previous level. For the ANFIS in Fig. 2.1 the following stands:

- $f_1 = 0.1*0.75 + 0.3*0.33 + 10 = 10.249$
- $f_2 = 0.25 + 0.1*0.67 + 5 = 5.317$
- $\overline{w_1} * f_1 = 0.4589 * 10.249 = 4.703$
- $\overline{w_2} * f_2 = 0.5411 * 5.317 = 2.877.$

Level 5–The Output Layer

The node of this level is a fixed node, which is labeled as 'Σ'. The output of this node is the overall system's output, and it is calculated adding all the input values. For the ANFIS in Fig. 2.1 is $\Sigma = \overline{w_1} * f_1 + \overline{w_2} * f_2 = 4.703 + 2.877 = 7.58$.

Chapter 3
Fuzzy Logic-Based Software Systems

Abstract Fuzzy logic was introduced by Lotfi A. Zadeh in 1965 as a methodology for computing with words and since then it is used extensively. It is an artificial intelligent technique that is ideal for successfully addressing, , the uncertainty, imprecision and vagueness that exist in many diverse scientific and technological areas. It is used in software systems to automatically predict, model, decide, diagnose, recommend etc.. In this chapter, several Fuzzy Logic-based software systems in the discipline of medicine, education, decision making and recommendation, natural language processing, automotive engineering and industry, heating, ventilation and air-conditioning, navigation, scheduling, network traffic and security are presented. This chapter aims to provide deep insights and valuable information about significant advances of Fuzzy Logic in software systems to researchers, scientists, educators, students, software engineers and developers.

3.1 Fuzzy Logic-Based Medical Systems

The development of expert systems in the medical field was begin at early 1970s. Examples of the first known medical expert systems are MYCIN [1], CADIAG [2], CASNET [3], and INTERNIST [4]. The main aim of medical expert systems is to process the data, which are related with symptoms, diagnostic tests, medical images, clinical analysis and medical knowledge, and diagnose or predict a disease in an accurate way. To this purpose Artificial Intelligence is very helpful. Especially, fuzzy logic plays a major role in the expert diagnostic medical systems. The important contribution of fuzzy logic in medicine had been anticipated by Zaheh in 1969, when he written in his published paper that "the complexity of biological systems may force us to alter in radical ways our traditional approaches to the analysis of such systems. Thus, we may have to accept as unavoidable a substantial degree of fuzziness in the description of the behaviour of biological systems as well as in their characterisation. This fuzziness, distasteful though it may be, is the price we have to pay for the ineffectiveness of precise mathematical techniques in dealing with systems comprising a very large number of interacting elements or involving a large number of variables in their decision trees" [5]. From then until today there

K. Chrysafiadi, *Fuzzy Logic-Based Software Systems*, Learning and Analytics in Intelligent Systems 34, https://doi.org/10.1007/978-3-031-44457-9_3

Fig. 3.1 Statistics concerning published studies in fuzzy-based medical systems

is an increased scientific interest in incorporating fuzzy logic and Fuzzy Inference systems in expert medical diagnostic systems. From 2012 until 13 August 2022, 7,114 scientific studies related to fuzzy medical systems were published. Also, the last five years, an exponential growth in the number of articles in medicine field that comprise fuzzy logic is observed (Fig. 3.1).

The main reasons for which research studies choose to embed fuzzy logic techniques in the expert medical systems, are the following:

1. Medical diagnosis and data deals with uncertainty and imprecision. Sources of the uncertainty and subjectivity in medicine are:
   - Patient's symptoms definition and description.
   - Patient's medical history description.
   - Analysis of laboratory and diagnostic tests.
   - Complexity of medical images.
   - Diseases classification.
   - The complexity of human biological and physio-pathological processes.
   - The huge amount of medical data and knowledge.
2. Furthermore, disease descriptions, medical data analysis and medical knowledge are expressed with linguistic terms.

Below, many cases of recently developed medical systems that uses fuzzy logic are presented. Fuzzy logic in medicine is used, mainly, as diagnostic tools or for medical image's analysis and segmentation. The developed fuzzy medical systems have been proved to be a useful and supportive tool to clinicians and doctors.

3.1.1 *Examples of Fuzzy Diagnostic Medical Systems*

In this sub-section several recent expert medical systems, which embeds fuzzy logic in order to enhance the diagnosis of a variety of diseases, are presented. These systems, the incorporated fuzzy algorithm or FIS that they use, their input and output are sum up in Table A.1 in Appendix A.

- The authors, in [6] uses fuzzy K-nearest neighbors (FKNN) algorithm for enhancing the prediction of lupus nephritis. Three medical sets were used for the validity of the proposed model. The model's application for lupus nephritis prediction proves that the proposed model is an effective clinical diagnosis method.
- In [7], a Mamdani FIS, which classifies automatically children with cerebral palsy according to their gait patterns, is presented. To construct the FIS, the authors had advised the related literature review. The input of the system includes 8 parameters, which are associated with kinematic data. These parameters are knee flexion, ankle dorsiflexion, foot progression angle, the range of motions of hip and knee in the sagittal plane, the value of peak knee flexion, the average and the maximum angular velocity of the knee during loading response and midstance. For each parameter, four Gaussian membership functions were defined. The fuzzy rules of the proposed system, which imitate the human behavior concerning the classification of gait patterns in children with cerebral palsy were presented in Table 3.1. The output of the proposed system is the category of gait pattern. For the fuzzy output representation four triangular fuzzy sets were defined: (i) true equinus, (ii) apparent equinus, (iii) jumb, (iv) crouch. To produce the crisp output, the system uses the middle of maxima defuzzification method. The proposed FIS was validated using data from 33 affected limbs of a clinical database. The system identified more accurately true equinus and crouch patterns (sensitivity and specificity of 100%). Jump gait patterns were identified with a sensitivity of 90% and specificity of 96%. While, it identified the apparent gait pattern with a sensitivity of 86% and specificity of 96%.
- In [8], the authors used a FIS to classify mammographs in density classes to assist breast cancer diagnosis. The proposed system takes as input 6 features of the density map of a mammogram and classifies the mammograph to four classes, which concerns its density. These features are: adipose, fibroglandurar, low density, medium density, high density, number of connected components. The membership function of the input is calculate dynamically using Fuzzy C-means algorithm.
- In [9], a fuzzy logic-based diagnostic algorithm for implantable cardioverter defibrillators is presented. The proposed algorithm is based on the heart rate variability analysis. Its input includes RR-interval (I), as extracted from raw intracardiac electrogram, and in addition two other features of heart rate variability: onset and instability. Each input variable is described by 10 fuzzy sets, which are defined by triangular membership functions. 15 fuzzy rules are applied over them to

Table 3.1 Fuzzy rules that classify gait pattern

	Input								Output
	Hip flexion	Knee flexon	Ankle dorsiflexion	Foot progression	Knee flex—range of motion	Maximum velocity of knee flexion	Average velocity of knee flexion	Peak knee flexion value	Gait pattern
Rule 1	Slightly positive	Strongly negative	Strongly negative	Moderately negative	Wide	Fast	Medium	Low	True equinus
Rule 2	Strongly positive	Slightly positive	Moderately negative	Strongly negative	Wide	Medium	Fast	High	Jump
Rule 3	Moderately positive	Slightly positive	Slightly negative	Moderately negative	Medium	Fast	Medium	High	Apparent equinus
Rule 4	Strongly positive	Strongly positive	Strongly positive	Slightly positive	Narrow	Slow	Slow	Medium	Crouch

define the value of the output, which includes 6 diagnostic categories: ventricular fibrillation, ventricular tachycardia, sinus tachycardia, detection artifacts and irregularities (including extrasystoles), atrial tachyarrhythmias and no tachycardia (i.e. normal sinus rhythm). The output membership functions of the proposed fuzzy logic-based algorithm are singletons. The fuzzy rules are applied with a product inference and a weighted linear combination of the results. Real life recordings of raw intracardiac electrograms, MIT-BIH Arrhythmia Database and MIT-BIH Supraventricular Arrhythmia Database were used for training, validation and testing of the algorithm. The proposed diagnostic algorithm was proved that has the ability to reduce the rate of occurrence of inappropriate therapies. Moreover, its computational complexity is low enough to be used in implantable medical devices.

- In [10] the authors present a hybrid model, which includes three classifier models of neural networks, type-1 and type-2 Mamdani FIS, as classification method for 2-lead cardiac arrhythmia. Heartbeats signals is given as input for classification three different classifiers: the 1st classifier is based on fuzzy KNN algorithm, the 2nd classifier includes a multilayer perceptron with gradient descent and momentum, and the 3rd classifier concerns a multilayer perceptron with scaled conjugate gradient backpropagation. For the classification process ten different classes of normal and arrhythmia signals are considered. The outputs from the classifiers are inputs to two FIS: a type-1 Mamdani FIS and a type-2 Mamdani FIS. In more details, the two FIS take as inputs 30 values: 10 inputs correspond to the output of the fuzzy KNN algorithm classifier, the next 10 inputs correspond to the second classifier, and the last 10 inputs correspond to the third classifier. The inputs are characterized by three fuzzy sets: 'Low_Classification', 'Medium_Classification' and 'High_Classification'. Their membership functions are trapezoidal. The systems' output is the integration of the classifiers' results. Therefore, the output of the two FIS concerns the degree of classification of a heartbeat signal to the ten defined classes. The output is characterized, also, by trapezoidal membership functions of the tree fuzzy sets 'Low_Classification', 'Medium_Classification' and 'High_Classification'. The output defuzzification method is centroid. The type-1 FIS uses 260 rules to combine the outputs of the classifiers regarding the membership of the fuzzy KNN algorithm and the activation of the other two classifiers. The type-2 FIS uses 90 fuzzy rules. Training and testing of the hybrid model were with the Massachusetts Institute of Technology and Beth Israel Hospital (MIT-BIH) arrhythmia database. The results showed that the proposed hybrid model succeeds better performance and classification rate concerning the cardiac arrhythmia classification problem.
- In [11] the authors proposed a fuzzy logic-based expert system for diagnosis and prediction of chronic kidney disease and evaluated its robustness against noisy data. They developed a Mamdani FIS, whose input contains 16 diagnostic parameters: age, systolic blood pressure, diastolic blood pressure, proteinuria, albuminuria, creatinine, glomerular filtration rate, blood urea nitrogen, parathyroid hormone, hemoglobin, calcium, phosphorus, polar height to pole in ultrasound, history of hypertension, history of diabetes, and family history of CKD. The

defined fuzzy sets of the input variables are presented in Table 3.2. The membership functions of the input variables are Gaussian, Trapezoidal, Triangular, and singleton. It was selected the membership function that suits better to each input variable. The developed system has 36 fuzzy rules, which are fired independently. The identification of the 16 diagnostic parameters, the definition of their membership functions and the creation of the fuzzy rules were based on the literature review, guidelines and consulting 18 nephrologists. The system's output is only CKD variable, which corresponds to the chronic kidney disease diagnosis. The designed FIS was evaluated using data extracted from 216 randomly selected medical records of patients with and without chronic kidney disease. The results showed that the proposed FIS predicts chronic kidney disease with 92.13% accuracy, 95.37% sensitivity and 88.88% specificity.

- The authors of [12] proposed a fuzzy ensemble feature selection methodology for knee osteoarthritis (KOA) prediction with special emphasis on the progression of joint space narrowing (JSN). Particular, a two level Mamdani FIS is employed to combine multiple feature importance scores in order to select the most informative features for KOA prediction. The 1st level of Mamdani FIS is used to define the feature scoring of three feature selection methods (filter, wrapper and embedded). Each one of these feature selection methods has two different categories. The results of each category constitute the input of the 1st level of Mamdani FIS.

Table 3.2 Fuzzy sets of 16 input variables

Input variable	No of fuzzy sets	Linguistic values
Age	3	Young, middle-aged, old
Systolic blood pressure	2	Normal, abnormal
Diastolic blood pressure	2	Normal, abnormal
Proteinuria	2	Normal, abnormal
Albuminuria	2	Normal, abnormal
Creatinine	2	Normal, abnormal
Glomerular filtration rate	3	Normal, before the disease, during the disease
Blood urea nitrogen	2	Normal, abnormal
Parathyroid hormone	2	Normal, abnormal
Hemoglobin	2	Normal, abnormal
Calcium	2	Normal, abnormal
Phosphorus	2	Normal, abnormal
Polar height to pole in ultrasound	3	Normal, before the disease, during the disease
History of hypertension	2	Low, high
History of diabetes	2	Low, high
Family history of chronic kidney disease	2	No, yes

Three fuzzy sets were defined for the description of the feature selection of each category of feature selection methods: 'Low', 'Medium', 'High'. The membership functions of 'Low' and 'High' are trapezoidal, while the membership function of 'Medium' is triangular. The output of the 1st level of Mamdani FIS is the feature importance for each feature selection method. It is described by the following five fuzzy sets: 'Very Low', 'Low', 'Medium', 'High', 'Very High'. The membership functions of 'Low' and 'High' are trapezoidal, while the membership functions of the rest fuzzy sets are triangular. The fuzzy set of the 1st level of Mamdani FIS includes 9 rules with the format "If Feature_Selection_Method1 is A and Feature_Selection_Method1 is B then Feature_Importance is C". The defuzzification values of the 1st level of Mamdani FIS for each feature selection method constitute the inputs of the 2nd level of Mamdani FIS. Therefore, the input of the 2nd level of Mamdani FIS includes the three following variables: FilterScoring, WrapperScoring, EmbeddedScoring. The linguistic values of these fuzzy variables are 'Low', 'Medium', 'High'. The membership functions of 'Low' and 'High' are trapezoidal, while the membership function of 'Medium' is triangular. The output of the 2nd level of Mamdani FIS is the aggregated feature importance. It is described by the following five fuzzy sets: 'Very Low', 'Low', 'Medium', 'High', 'Very High'. The membership functions of 'Low' and 'High' are trapezoidal, while the membership functions of the rest fuzzy sets are triangular. The fuzzy set of the 2nd level of Mamdani FIS includes 27 rules with the format "If FilterScoring is A and WrapperScoring is B and EmbeddedScoring is C then Feature_Importance is D". The proposed fuzzy logic-based methodology was evaluated using data from the Osteoarthritis Initiative database. The evaluation results showed that the proposed method presented a better and more stable performance in contrast to other competitive feature selection methods.

- In [13] an effective and efficient diagnosis system, which uses an adaptive fuzzy k-nearest neighbor approach, for Parkinson's disease is presented. The neighborhood size k and the fuzzy strength parameter m in FKNN classifier are adaptively specified. The proposed approach was compared to other related methods. The results showed that the proposed system achieves the highest classification accuracy.
- In [14] the authors used the Fuzzy Support Vector (FSVM) Machine for detecting diabetes mellitus. Initially, the researchers considered the following 8 diagnostic characteristics of a patient from the National Institute of Diabetes and Digestive and Kidney Diseases datasets: number of pregnancies, plasma glucose concentration, blood pressure, skin fold thickness, insulin, body mass index, diabetes pedigree function, age. After the application of FSVM, two only features were considered as valuable. These features are: plasma glucose concentration and body mass index. For the representation of them the three following fuzzy sets were used: 'low', 'medium', 'high'. The membership functions for the fuzzy sets of plasma glucose concentration are trapezoidal. The membership function for fuzzy sets 'low' and 'high' of body mass index are trapezoidal, while the membership function for the fuzzy set 'medium' is triangular. Also, nine fuzzy rules for diabetes mellitus diagnosis based on plasma glucose concentration and body mass index

were generated. The evaluation results of the proposed methodology had 89.02% accuracy in predicting patients with diabetes mellitus.

- The study in [15] concerns a fuzzy rule-based clinical system for the automatic detection of Coronary Heart Disease. The authors designed a Mamdani FIS, which has 7 input parameters, which are cholesterol, blood pressure, physical activity, smoking, age, BMI, and diabetes. For the description of cholesterol four fuzzy sets (low, medium, high and very high) are used. The membership functions of "normal" and "very high" fuzzy sets are trapezoidal, while the membership functions of "Medium" and "High" fuzzy sets are triangular. For the description of blood pressure three fuzzy sets (normal, medium, high) are used. The membership functions of "normal" and "high" fuzzy sets are trapezoidal, while the membership function of "Medium" fuzzy set is triangular. 'Physical activity' indicates how regular the patient exercises. It has two linguistic values (false or true), which described by trapezoidal membership functions. Similarly, the 'smoking' input variable takes the linguistic values true and false. For the description of age four fuzzy sets (young, middle age, old and very old) are used. The membership functions of "young" and "very old" fuzzy sets are trapezoidal, while the membership functions of "middle age" and "old" fuzzy sets are triangular. For the description of BMI three fuzzy sets (normal, over-weight, obese) are used. The membership functions of "normal" and "obese" fuzzy sets are trapezoidal, while the membership function of "over-weight" fuzzy set is triangular. For the description of diabetes two fuzzy sets (normal and diabetic) are used. Their membership functions are triangular. The system's output concerns the presence of heart disease in the patient. It is described by the following linguistic values: healthy, early stage, and advanced stage. The membership functions of "healthy" and "advanced stage" fuzzy sets are trapezoidal, while the membership function of "early stage" fuzzy set is triangular. The system incorporates 44 fuzzy rules. The format of these rules is "If cholesterol is A and blood pressure is B and physical activity is C and smoking is D and age is E and BMI is F and diabetes is G then disease_classification is Z". For the system's development patient clinical data of databases at the University of California, Irvine, were concerned.
- In [16] the authors constructed a Fuzzy Cognitive Map (FCM) or diagnosing rheumatic-musculoskeletal disease. For the construction of the FCM a Mmadani FIS was used. Particularly, 28 diagnostic parameters & their degree of relationships are the system's input. The 28 diagnostic parameters are joint stiffness, joint deformity, joint pain, joint swellings, dry eyes, mouth sores, genital sores, dry mouth, fever, unsteady gait, skin rash, weight loss, twitching muscles, fatigue, paraesthesia, muscle weakness, sleep disturbances, back pain, headache, hotness of the joints, foamy urine, vomiting, diarrhoea, chest pain, redness of eyes, breathlessness, skin nodules, colour changes. Each parameter is described by five triangular fuzzy sets (Absent, Mild, Moderate, Severe and Very Severe). The degree of a relationship between two diagnostic parameters is described by the following five triangular fuzzy sets: no relationship, weak relationship, strong relationship, and very strong relationship. The degree of relationships was asked to be defined

by experts and constitute the fuzzy rules. The system's output considers the diagnosis of rheumatic-musculoskeletal disease and is described by the following five triangular fuzzy sets: Absent, Mild, Moderate, Severe and Very Severe. The defuzzification of the output is performed by using the centre of gravity method. The system's evaluation results described by five triangular fuzzy sets (Absent, Mild, Moderate, Severe and Very Severe). The system's evaluation results showed 87% accuracy, 90% sensitivity, and 80% specificity of the proposed diagnostic approach.

- In [17] a fuzzy logic-based expert system for cervical neoplasia (CN) precursor injuries diagnosis is presented. The proposed expert system includes three stages: the stage of risk diagnosis, the stage of cytological image interpretation, and the stage of injuries classification. Two Mamdani FIS are used in the first and third stage. Particularly, in the first stage a FIS is used to measure the risk for contracting CN according to the patient's clinical records. The FIS takes 7 fuzzy variables as inputs (age, onset of sexual activity, sexual couples, number of pregnancies, first pregnancy age, sexually transmitted diseases, cervical lesions) and gives a fuzzy variable as output (risk). For the description of 'age', 'onset of sexual activity' and 'first pregnancy age' three trapezoidal fuzzy sets are used (young, adult, mature). For the description of 'sexual couples' three trapezoidal fuzzy sets are used (few, many, unusual). For the description of 'number of pregnancies' two trapezoidal fuzzy sets are used (few, many). For the description of 'sexually transmitted diseases' and 'cervical lesions' two triangular fuzzy sets are used (no, yes). The fuzzy output can take two linguistic values: negative and positive. Their membership functions are triangular. For the defuzzification of the output the centroid method is used. The FIS for risk diagnosis uses 360 fuzzy rules. The FIS of the 3rd stage links the obtained results in the fuzzy model of the first stage with the results of the images processing (2nd stage) in order to classify injuries. The fuzzy system's input is nucleus dimension, nuclei staining, homogeneous nucleation, nucleolus presence and risk. For the description of 'nucleus dimension' three trapezoidal fuzzy sets are used (small, medium, large). For the description of 'nuclei staining' three trapezoidal fuzzy sets are used (clear, media, dark). For the description of 'homogeneous nucleation' two trapezoidal fuzzy sets are used (regular, irregular). For the description of 'nucleolus presence' three fuzzy sets are used (null, few, many). The membership function of 'null' is triangular, while the membership functions of 'few' and 'many' are trapezoidal. For the description of 'risk' two triangular fuzzy sets are used (negative, positive). The FIS's output 'Injury resolution' is described by three trapezoidal fuzzy sets (no alterations, atypical glandular cells, malign positive). The proposed system was validated by testing 21 already diagnosed cases. The system had 100% effectiveness during the validation process.
- In [18], a fuzzy rule-based expert system that supports doctors in chronic kidney disease diagnosis is presented. The authors developed a Mamdani FIS that takes as input seven characteristics of the patient and detects the stage of chronic kidney disease. The input variables are nephron functionality, blood sugar, diastolic blood pressure, systolic blood pressure, age, body mass index, and smoke. For the

description of 'nephron functionality' three trapezoidal fuzzy sets are used (very risky, moderately risky, safe zone). For the description of 'blood sugar' three trapezoidal fuzzy sets are used (safe, borderline, high). For the description of 'diastolic blood pressure' and 'systolic blood pressure' four trapezoidal fuzzy sets are used (low, medium, high, extremely high). For the description of 'age' three trapezoidal fuzzy sets are used (young, mid-age, old). For the description of 'body mass index' four trapezoidal fuzzy sets are used (underweight, normal weight, overweight, obese). Finally, for the description of 'smoke' three trapezoidal fuzzy sets are used (low, medium, high). The system applies 83 fuzzy rules over the input fuzzy variables. The format of these rules is "if nephron functionality is A and blood sugar is B and diastolic blood pressure is C and systolic blood pressure is D and age is E and body mass index is F and smoke is G then patient's kidney is Z". Therefore, the system's output is the patient's condition considering chronic kidney disease. The output is described by 6 fuzzy sets (healthy, concerning, very concerning, sick, very sick, extremely sick). 'Healthy' and 'extremely sick' are defined by trapezoidal membership functions, while the rest 4 output fuzzy sets are defined by triangular membership functions. For the defuzzification of the output, the centroid method is used. The proposed fuzzy rule-based expert system is developed with the help of clinical practice guidelines, database, and the knowledge of a team of experts. For the evaluation of the proposed expert system, 80 tests were performed. The generated output was compared with expected output for each one of the 80 tests, and it is observed that the system succeeds in 93.75% of the tests.

- In [19], the authors developed a fuzzy expert system to diagnose the hypertension risk for different patients based on a set of symptoms and rules. The operation of the system is based on a neuro fuzzy system for the set of symptoms and rules using three different types of learning algorithms which are Levenberg–Marquardt (LM), Gradient Descent (GD) and Bayesian Resolution (BR) based learning functions. The input parameters are age body mass index, blood pressure, and heart rate. Their linguistic variables and corresponding membership functions were defined taking into account experts' knowledge and advises, statistical analysis of patient data and literature review. For the description of 'age' three triangular fuzzy sets are used (young, middle, old). For the description of 'body mass index' and 'heart rate' three triangular fuzzy sets are used (low, medium, high). For the description of 'blood pressure' three fuzzy sets are used (low, medium, high). The membership functions of 'low' and 'medium' are triangular, while the membership function of 'high' is trapezoidal. The system's output is the risk of hypertension, which described by three triangular fuzzy sets (less, moderate and severe). The set of fuzzy rules includes 9 rules. The system's linguistic variables, their membership functions and the fuzzy rules are used to construct the neuro fuzzy system.
- In [20] an intelligent medical diagnostic system, which embeds a two-layered fuzzy inference system, to diagnose renal cancer is described. The system's first layer is a Mamdani FIS that takes decision about whether the patient is having

renal cancer or not. The system's input includes 4 parameters (smoking, dialysis, occupational exposure, genetic or hereditary), whose linguistic values are described by Gaussian membership functions. The system's output is a fuzzy variable, which refers to the existence or not of renal cancer and is described by two Gaussian fuzzy sets (cancer present, no cancer). The rules in the first layer inference system are 36. The second layer of the inference system, which includes an ANFIS, detects the current stage of suffering patients. The following 7 parameters are the ANFIS's input: blood in urine, red blood cell count, flank pain, tumor size, Von Hippel-Lindau gene, high blood pressure, and trichloroethylene exposure. The output is a fuzzy variable (cancer stage) that is described by four fuzzy sets (minor, moderate, serious, critical). All input parameters and the system's output are described by Gaussian membership functions. The second layer has 864 rules in the IF–THEN form.

- In [21] a hierarchical fuzzy inference system for cardiovascular risk level prediction. The proposed system consists of three layers. Each layer has its own FIS. The output of a FIS is the input in the FIS of the next layer. Particularly, the input of the FIS of the 1st layer is the fuzzy variable 'blood oxygen', whose linguistic values are 'critical', 'low' and 'normal', and the fuzzy variable 'color lips', which represents the nuance of the lips and whose linguistic values are 'regular', 'altered' and 'purplish', and the fuzzy variable 'color lips'. The input of the FIS of the 2st layer is the output of the 1st layer (y1 with linguistic values: low, medium, and high) FIS and the fuzzy variable 'breathing rate', whose linguistic values are 'bradypnea', 'normal' and 'tachypnea'. The input of the FIS of the 3rd layer is the output of the 2nd layer FIS (y2 with linguistic values: low, medium, and high) and the fuzzy variable 'heart rate', whose linguistic values are 'bradycardia', 'normal' and 'tachycardia'. The output of the FIS of the 3rd layer is the risk level, which is defined by four linguistic terms to be intended as class labels representing different levels of cardiovascular risk: Low, Medium and High and Very High. The output of the FIS of the 3rd layer is the output of the overall system. Memberships functions of all input and output variables are triangular. The designed hierarchical fuzzy inference system has 81 fuzzy rules, which were defined concerning guidelines and physicians' knowledge.
- In [22] the authors present a fuzzy expert system (FES) for heart diseases diagnosis. The main part of the proposed expert system is a Mamdani FIS, which has three fuzzy variables as input and a fuzzy variable as output. Particularly, the three input fuzzy variables are systolic blood pressure, body mass index and triglycerides. For the description of 'systolic blood pressure' seven triangular fuzzy sets are used (optimal, normal, normally high, light, moderate, severe, isolated). For the description of 'body mass index' five triangular fuzzy sets are used (normal, overweight, grade 1, grade 2, grade 3). For the description of 'triglycerides' four triangular fuzzy sets are used (normal, medium, high, very high). The system's output concerns the risk of having a cardiovascular disease. It is described by five triangular fuzzy sets: very low risk, low risk, medium risk, high risk, and very high risk. The proposed expert system includes, also decision tree-based solution for the higher number of rules in order to increase the degree of efficiency. For the

validation and the evaluation of the proposed system clinical data from Cleveland heart diseases Database were used.

- In [23] a fuzzy logic-based expert system for lung cancer diagnosis is presented. The proposed system is based on a Takagi–Sugeno FIS. Input source of the system relies on literature studies and interviews with the doctors or physicians. Five fuzzy variables are used as input: anamnesis, doubling time, the degree of the smokers, age and performance status. For the description of 'anamnesis' and 'the degree of the smokers' three linguistic values are used (light, medium, heavy). The linguistic values of the variable 'age' are 'young', 'middle age', 'old'. The linguistic values of the variable 'performance status' are 'very weak', 'weak', 'normal'. The linguistic values of the variable 'doubling time' are 'normal, 'middle', 'high'. Regarding design rules, inside each fuzzy variable, the value is combined each other with their linguistic attribute according to the regulation or rules made by doctors or physicians. The set of rules includes 62 fuzzy rules. The output of the system is a diagnose that concerns the existence of cancer or not. More specifically, the output can be a value lower that 30, which indicates that the patient has not cancer, or a value between 30 to 70, which means that the patient has benign cancer, or a value over 70, which indicated that the patient has malign cancer.
- In [24] an expert system, that uses a Mamdani FIS, for diagnosis of the inflammatory diseases of the knee joint was presented. All the input parameters required are collected with Survey methods, interviews, research papers, books and expert doctors. They concerned symptoms for three knee inflammatory diseases: osteoarthritis, rheumatoid arthritis, and osteonecrosis. Particularly, for osteoarthritis, the following five input parameters are concerned: pain, age, swelling, deformity, and restriction. The input variables for rheumatoid arthritis are pain, age, ESR, bilateral swelling, and bilateral tenderness. Also, the input variables for osteonecrosis are pain, swelling, medial joint line tenderness, and restriction. All input variables are described by thee fuzzy sets: low, medium, high. The membership function of 'low' and 'high' are triangular, while the membership function of 'medium' is trapezoidal. The output of the system includes three fuzzy variables, which correspond to inflammatory diseases osteoarthritis, rheumatoid arthritis and osteonecrosis. The fuzzy variable 'osteoarthritis' takes the following three linguistic values: early, moderate, severe. The linguistic values for the fuzzy variable 'rheumatoid arthritis' are sero−ve and sero+ve. The linguistic values for the fuzzy variable 'osteonecrosis' are stage1 and stage2. The membership functions of the output variables are trapezoidal and triangular. Furthermore, or the osteoarthritis, there are 143 rules; for the rheumatoid arthritis there are 152 rules and for the osteonecrosis there are 68 rules. The fuzzy rules were constructed with experts' help. The defuzzification of the output is performed by centroid method.
- In [25], the authors describe a pre-clinical medical diagnosis system for COVID-19 and other respiratory conditions with similar symptoms: allergies, common cold and influenza. The FIS is based on the Mamdani model. The system has 9 inputs: (1) fever, which has three linguistic values: no, mild, severe, (2) headache, which has two linguistic values: no and yes, (3) myalgia, which has two linguistic

values: no and yes, (4) fatigue, which has two linguistic values: no and yes, (5) nasal congestion, which has two linguistic values: no and yes, (6) sneezing, which has two linguistic values: no and yes, (7) sore throat, which has three linguistic values: no, mild, severe, (8) difficulty breathing, which has two linguistic values: no and yes, and (9) rhinorrhea, which has two linguistic values: no and yes. The membership functions of the input variables are sigmoidal or generalized bell. Also, the system has 4 outputs: $\mu_C(x)$, $\mu_A(x)$, $\mu_{CC}(x)$, and $\mu_I(x)$, which denote the degree to which x belongs to each class, or diagnosis, of the set (1) COVID-19, (2) allergies, (3) common cold and (4) influenza, respectively. Each output variable has four possible linguistic values: unlikely, possible, probable and very probable. The final diagnosis is calculated as the maximum of each of the FIS outputs. The system includes 10 fuzzy if–then rules, which used to classify the different combinations of symptoms for the conditions studied (COVID-19, allergies, common cold, and influenza). Symptom data, the input definition, the membership functions and the set of fuzzy rules were carried out empirically based on the analysis of literature previously reported.

- In [26], the authors propose a predictive model for heart disease diagnosis using a fuzzy rule-based approach with decision tree. The presented fuzzy system is a Mamdani FIS, which has 8 fuzzy variables as input and one fuzzy variable as output. Particularly, the input variables concern eight attributes of a patient with heart disease. The output concerns the heart disease diagnosis. The membership functions of the system are triangular and trapezoidal. The constructed FIS includes fuzzy rules, which are extracted by a decision tree. The decision tree is built considering the linguistic values of the fuzzification of the input variables. It consists of a root node representing the top node of the tree, leaf node representing the class label, internal nodes representing the attribute, and branches representing each possible value of the attribute node from which they originate. For the construction and evaluation of the proposed system a dataset that was collected from UCI machine learning repository, was used. This dataset provided the results of various medical tests performed on a patient and also the diagnosis of presence or absence of heart disease on the patients. It contained 14 attributes and 303 examples of patient. The accuracy of the proposed system was 88%.
- In [27], a cardiovascular diagnosis system using fuzzy logic approach is presented. The system's aim is the early detection of heart diseases, depending on cardiovascular risk factors. The deigned system has 10 input variables, which concern patients' clinical data. The input variables are: heart rate, patient age, systolic blood pressure, diastolic blood pressure, sex, chest pain, diabetes, smoking, electrocardiogram and total cholesterol. The linguistic values of 'heart rate' are normal, tachycardia and bradycardia. The linguistic values of 'patient age' are young, medium, old and very old. The linguistic values of 'systolic blood pressure' and 'diastolic blood pressure' are optimal, normal, normally high, light, moderate, severe and isolated. The linguistic values of sex' are woman and man. The linguistic values of 'chest pain' are no pain, typical angor, atypical angor,

no anginal pain and asymptomatic. The linguistic values of 'diabetes' are non-diabetic, prediabetic and diabetic. The linguistic values of 'smoking' are non-smoking and smoking. The linguistic values of 'electrocardiogram' are normal, ST-T and hypertrophy. The linguistic values of 'total cholesterol' are normal, medium and high. The system' output is a fuzzy variable, which represents the percentage risk of having a cardiovascular disease. It is described by the following five fuzzy stets: very low risk, low risk, medium risk, high risk, and very high risk. The membership function of input and output variables are triangular. Furthermore, the system incorporates a set of fuzzy rules, which embeds all the necessary knowledge of expertise, experience and medical experts. The fuzzy rules provide the desired decisions concerning detection of heart diseases. For the output defuzzification to a crisp number, the centroid method is used.

- In [28], the authors propose a fuzzy inference system for osteoporosis disease diagnosis. The input of the system is sex, age, body mass index, amount of paint, heredity, calcium deficiency, physical stress, years since menopause and consumption alcohol or cigarette. For the description of 'age' three trapezoidal fuzzy sets (young, middle and old) are used. For the description of 'body mass index' four trapezoidal fuzzy sets (underweight, normal, overweight and obese) are used. For the description of 'amount of paint' three trapezoidal fuzzy sets (mild, moderate and severe) are used. For the description of 'calcium deficiency' three trapezoidal fuzzy sets (low diet, medium diet and heavy diet) are used. For the description of 'physical stress' two trapezoidal fuzzy sets (medium and high) are used. For the description of 'years since menopause' one triangular fuzzy set is used. For the description of 'consumption alcohol or cigarette' two triangular fuzzy sets (low and high) are used. The rest input variables (sex and heredity) are crisp. The system's output concerns the osteoporosis detection and takes the three following linguistic values: normal, osteopenia, osteoporosis. The memberships functions of the output fuzzy sets are trapezoidal. The system applies if–then fuzzy rules, which is based on experts' knowledge, to give the output result. The proposed fuzzy inference system is combined with x-ray image processing to take a final decision about the osteoporosis disease detection.
- In [29] a Mamdani FIS for diagnosis of anemia disease is presented. The system's input includes the content of serum iron in the blood (SFe), the total iron binding capacity of blood serum (TFCS), the serum ferritin content (FS) and the reticulocyte count (RTC). The input variables are described by three triangular fuzzy sets (reduced, norm, increased). The system decides about the type of anemia that the patient deals with. Therefore, the output of the system is a fuzzy variable, which can take the following five linguistic values: iron-deficiency anemia, anemia associated with impaired porphyrin synthesis, absence of anemia, anemia in chronic diseases, thalassemia. The membership functions of the output fuzzy sets are triangular, too. The fuzzy rules of the system have the format "if SFe is A and TFCS is B and FS is C and RTC is D then anemia is Z".
- In [30] a system for coronary heart disease prediction is presented. The system is based on Mamdani FIS. It includes 6 continuous input variables (age, total cholesterol, LDL cholesterol, HDL cholesterol, systolic blood pressure, and diastolic

blood pressure) and 3 categorical input variables (sex, smoking, and diabetes). For the definition of the continuous data, triangular and trapezoidal fuzzy sets are used. In particular, the triangular fuzzy sets 'young', 'middle' and 'old' are used to describe the age. For the description of 'total cholesterol', 'LDL cholesterol', and 'HDL cholesterol' the following three triangular fuzzy sets are used: low, middle and high. The input variables 'systolic blood pressure' and 'diastolic blood pressure' takes five linguistic values (very low, low, middle, high, very high). The membership functions of 'very low' and 'very high' are trapezoidal, while the membership functions of the rest fuzzy sets are triangular. The system uses a set of fuzzy rules to perform inference. The rules were generated using a decision tree technique. The system's output is the coronary heart disease risk level. It takes two linguistic values: low risk and high risk. The membership functions of the output fuzzy sets are triangular. The output defuzzification is performed by using the center of gravity method. For the proposed system's design and evaluation, datasets derived from the Korean National Health and Nutrition Examination Survey VI were utilized.

- In [31] the authors introduced a hybrid diagnose strategy for quickly and accurately diagnose of COVID-19 cases. The proposed diagnose strategy consists of three phases: (i) pre-processing phase, in which patients' data of datasets are filtered using data mining techniques and selected features are extracted, (ii) feature ranking phase, in which the selected features are ranked. and (iii) classification phase, in which the ranked features are used to derive the classification model that can classify new persons to decide whether they are infected or not. In the classification model a FIS is used. Particularly, FIS takes as inputs the fuzzy variables 'White Blood Cell', 'Lymphocyte', 'Monocytes', and 'Locate Dehydrogenase'. The output of the FIS is the COVID-diagnosis value. All input and output variables of the FIS can take the following three linguistic values: 'Low', 'Medium', 'High'. The membership functions of the fuzzy sets 'Low' and 'High' are trapezoidal, while the membership function of the fuzzy set 'Medium' is triangular. The system decides about the output value applying a set of 81 fuzzy rules. The rules have the format "If White Blood Cell is A and Lymphocyte is B and Monocytes is C and Locate Dehydrogenase is D then output is E", where A, B, C, D, E are 'Low', 'Medium' or 'High'. The output defuzzification is performed using the Center of Gravity (COG) method. At the last step of the proposed diagnosis system, the crisp output of the FIS is combined with the output of a Deep Neural Network and the system classifies the input case as patient or normal.

3.1.2 Examples of Fuzzy Medical Image Analysis and/or Segmentation

An important medical tool for disease prediction and diagnosis are medical images. Medical images include Computed tomography (CT) images, magnetic resonance images (MRI), magnetic resonance angiography (MRA), X-ray, mammography etc.

An important and complex task of medical image processing and analysis is image segmentation. Segmentation is the process of classification of the pixel to understand the information in the images. Therefore, it is crucial for accurate diagnosis. However, due to the medical image's high complexity and noise, segmentation effectiveness is limited. is the most complex and important task in the field of medical image processing and analysis, as it is linked to disease diagnosis accuracy. Many researchers use fuzzy logic in the image segmentation process to make it more accurate. Most cases include medical image analysis for cancer detection. In this sub-section several recently published fuzzy logic-based methodologies of medical image analysis and/or segmentation are presented. These published studies, the incorporated fuzzy algorithm that they use, their input and output are sum up in Table A.2 in Appendix A.

- In [32] the authors introduced a new approach in tumor detection using FCM algorithm. The input data is the set of pixels which are able to provide the information required to perform the mass segmentation by fixing two clusters only. The Gray Level Occurrence Matrix (GLCM) is used to extract the texture features for getting the optimal threshold. System decides based on contrast, energy, homogeneity, entropy and correlation and uses Breast Cancer and mammograms digital databases to be fed with material for evaluation. The output is a boundary produced by the system. The performance of the proposed method is evaluated by specificity, sensitivity and accuracy and the results obtained from experimentations show a good efficiency at the different measures used, in favor of the method proposed.
- In [33] the authors present an efficient framework for processing of mammogram images and introduces an algorithm for segmentation of the images to detect masses. They propose a fuzzy enhancement technique in the pre-processing step of mammogram images for contrast enhancement. The proposed fuzzy logic-based technique consists of fuzzification, modification of the membership function, and defuzzification. In the fuzzification process the gray-scale values of a mammogram image are converted to fuzzy values using membership functions. Then, the membership values resulting from the previous stage are modified using an appropriate fuzzy system, which generates new membership values. For contrast enhancement, the dark pixels will be darker, and the bright pixel will be brighter, but the pixels with moderate intensity values are not converted. Finally, the new gray-scale values are transformed from the modified fuzzy values back to the gray-scale domain by employing membership values to know the degree of brightness or darkness of pixels in the image. The proposed fuzzy image enhancement technique was analyzed and compared with conventional techniques and local contrast metrics. The comparison showed an outstanding performance of the proposed technique from the visual
- In [34] an algorithm based on an interval-valued intuitionistic fuzzy set is presented for efficiently fusing multimodal medical images. Multimodal medical image fusion is the process of combining two multimodal medical images to increase the quality and to extract maximum information from the output image for

better treatment and precise diagnosis. System receives input of medical images of all kinds, including MRI, x-ray scanners, digital medical imaging formats. The system calculates the optimum value computed for the image using the entropy Equation. System finds the fuzzified interval-valued intuitionistic fuzzy set image with the optimum value and a decision map (DM) is constructed that determines the combination of pixels of two certain images. Finally, a new decision map (NDM) is created by refining DM with consistency verification by using the majority filter and a crisp image is constructed by defuzzification of the fused image.

- In [35] the authors propose a segmentation method of medical images that is based on intuitionistic possibilistic fuzzy c-mean clustering. The proposed clustering approach applied for the clustering of mammogram images for breast cancer detection. The proposed system was designed in four stages: (i) pre-processing, (ii) main segmentation, (iii) statistical extraction, and (iv) final classification. The aim of combining the possibilistic fuzzy c-mean with intuitionistic fuzzy c-mean is to reduce the number of iterations to help with minimizing execution time. Possibilistic clustering approaches attempt to decrease the membership degree of noisy data, whereas the Intuitionist Possibilistic fuzzy clustering (IPFCM) approach assigns membership and non-membership degrees with hesitation degree. For the system's design and evaluation, a total of 320 digital mammogram images of the Mammography Image Analysis Society dataset were concerned. The proposed method has been compared with other available fuzzy clustering methods and the results showed that the proposed approach is highly effective with clustering and also with classification of breast cancer.
- In [36] a novel segmentation approach of magnetic resonance images (MRI), which combines the constrained fish school optimization (SCFSO) algorithm and interval type-II fuzzy clustering for the demarcation of the tumor and nontumor portions, is presented. The novel approach resolves the erroneous prediction of anomalies present in various topographical locations in brain subjects of MRI modality. The proposed approach consists of the following steps: (i) the preprocessing of input images, (ii) the application of SCFSO method to the preprocessed image to calculate the optimal cluster position, (iii) interval fuzzy clustering process and, (iv) demarcation of the tumor and nontumor portion. Therefore, the interval type-II fuzzy clustering method takes as input the optimal cluster boundaries, which were determined by SCFSO and give as output the tutor and non-tutor regions of MRI. The optimal cluster position obtained from SCFSO is utilized for identifying the cluster center by interval type-II method. The upper membership function and lower membership of the IT2FLS are used to frame the optimal cluster. SCFSO is a population-based search algorithm inspired in the behavior of swimming fishes in a school that expands and contracts while looking for food [37]. The proposed technique was applied to BRATS-SICAS dataset. The MRI segmentation outcomes of the proposed approach were very promising and showed that the time taken for performing the delineation of tumors and other regions was very less when compared to other conventional techniques.

- In [38], a new Internet of Things (IoT) based predictive modelling by using fuzzy cluster-based augmentation is presented. It uses classification for predicting the lung cancer disease. Here, Fuzzy C-Means Clustering algorithm is used to categorize the transitional region features from the feature of lung cancer image. In this work, Otsu [39] thresholding method is used for extracting the transition region from lung cancer image. The Fuzzy C-Means Clustering algorithm is used for segmentation which incorporates the transitional region-based feature extraction, morphological cleaning and thinning operation from the feature of lung cancer image. This work considered 113 lung images with high resolution Computed tomography (CT) medical images of dimension of 512 X 512 pixels in each slice. Generally, the left and right lungs are segmented in this work from the chest CT scan lung images which are available in the database by using the standard Fuzzy C-Means Clustering method-based segmentation process. In addition, other data such as blood pressure, heartbeat, sugar level details are collected through IoT device and also considered for making decision. This data is given as input to the system in the form of text and numbers. The evaluation results proved that the proposed model is able to achieve better accuracy when it is compared with other existing prediction model.
- In [40], the authors propose a rule-based fuzzy method for image segmentation of ultrasound image for the detection of breast cancer. After preprocessing of the ultrasound image, fuzzy clustering method is used to segment it into number of clusters. In the fuzzy clustering method initially number of clusters is assigned. Based on the number of clusters, membership function is generated. Using that membership function, calculate the centroid for each cluster. After calculating the centroid value, update the membership function and the centroid for every time. Numbers of pixels are grouped to make the cluster. The pixels which are belonging to one cluster may also belong to another cluster. After segmenting the image into set of clusters fuzzy level set algorithm is applied for more accurate detection of edges in the tumor region.
- In [41], the authors propose a system to help in the initial stage detection of lung nodules by classifying the existence of abnormalities in the computed tomography (CT) scan image. The detection method includes a neuro-fuzzy algorithm in the final step. Particularly, initially, CT images enhancement is performed; then image segmentation is applied; then several features from the enhanced segmented image are extracted; and finally, these extracted features are passed over the neural network to obtain the reality of the input lung image, whether it is a normal or abnormal lung image.
- In [42] a quadratic polynomial guided fuzzy C-means algorithm for medical image segmentation is proposed. The presented method distinguishes better the image's weak edge regions and has a degree of noise resistance, resulting in a membership matrix with less ambiguity and a more reliable segmentation result. For the evaluation of the proposed method, three medical datasets, which include mainly lung medical images, were used.
- In [43] the authors present a hybrid approach, which included fuzzy c-means algorithm and a fourth order partial differential equation based nonlinear filter,

for segmentation of microscopic biopsy images in presence of Poisson noise. The proposed approach succeeds to perform image segmentation and, simultaneously, to handle the noise present in the image. Particularly, it handles effectively the segmentation problem of blocky artifacts while achieving good tradeoff between Poisson noise removals and edge preservation of the microscopic biopsy images during segmentation process for cancer detection from cells. The presented method was tested on breast cancer microscopic biopsy data set.

- In [44] the authors present a method for Glioma brain tumor detection and segmentation. The proposed method uses Adaptive Neuro Fuzzy Inference System (ANFIS) approach to classify a brain image into normal and Glioma brain image. The presented ANFIS takes as input the texture features, which are extracted from the fused brain image, and classify the brain image as glioma image or non-glioma image. The fused image is derived by using the Non-Sub sampled Contourlet Transform [45]. For the evaluation of the proposed method the Brain Tumor image Segmentation challenge open access dataset was used.
- In [46] a novel level set method, which is based on an enhanced Fuzzy C-Means algorithm [47] and lattice Boltzmann method, to address aforementioned challenges for reliable and automatic brain tumor segmentation is presented. The proposed method includes the following four major steps: (i) superpixel segmentation [48], which is modified to be suitable for gray images segmentation of brain images, (ii) histogram of segmented image is calculated by superpixel approach. In this level, superpixel fuzzy clustering is performed through the enhanced Fuzzy C-mean algorithm, (iii) energy formulation is defined after obtaining the relevant level set formulas by using gradient descent method, and (iv) the lattice Lattice Boltzmann approach [49] is used. The combination of superpixel segmentation of fuzzy clustering handles the problem of the spatial information of pixels, which is caused by the simple Fuzzy C-Means algorithm.
- In [50], the authors describe a hybrid intelligent approach for identifying the pulmonary nodules in computed tomography (CT) images of lungs. The proposed approach consists of five fundamental building phases called image pre-processing, image segmentation, nodule candidate detection, feature extraction, and classification, respectively. Fuzzy logic is used in image pre-processing phase for image enhancement and in the image segmentation phase. Particularly, the proposed approach utilizes a type-II fuzzy algorithm to improve the quality of raw CT images. Also, a novel segmentation algorithm based on fuzzy c-means clustering is used to segment the enhanced CT images and achieve another representation of lung regions through an optimization methodology.
- In [51] a novel approach for analyzing magnetic resonance abdominal images and detecting the lumen and the aortic wall, is presented. The method combines different algorithms in two stages to improve the detection and the segmentation. In a first stage, fuzzy C-means algorithm and morphological image analysis are used to detect and segment the lumen. In the second stage, a graph cut algorithm is used to segment the aortic wall.
- In [52] the authors used a Fuzzy Inference System (FIS) in a method for iris localization and pupil detection in human eye for automated cataract detection.

The proposed FIS is used for the extraction of edges from eye images. FIS takes as inputs the vertical detail component and the horizontal detail component of the gray scale eye image, which are derived after applying Gaussian filtering. Based on these input variables a crisp output is obtained to determine the effective location of the edges based on the following two fuzzy rules: (1) "If the vertical detail component is zero and the horizontal detail component is zero then the location of the edges is white", (2) If the vertical detail component is not zero or the horizontal detail component is not zero then the location of the edges is black".

- In [53] the authors use ANFIS to classify brain tumor to abnormal and normal. The proposed ANFIS is part of a whole system for the detection and diagnosis of brain tumor that includes analysis of brain medical images.
- In [54] a clustering algorithm, which is based in fuzzy C-means algorithm, for MR brain image segmentation is described. The proposed algorithm combines fuzzy C-means algorithm with a multitask learning strategy to extract common knowledge from different but related MR brain image segmentation tasks. Furthermore, it avoids the negative effects caused by noisy data that exist in some MR images.
- In [55] the authors propose an Adaptive Neuro-Fuzzy Inference System (ANFIS) as a key rule for automatic COVID-19 detection from chest X-ray images based on the characteristics derived by texture analysis using gray level co-occurrence matrix (GLCM) technique. The aim is to construct an automated system to help the radiologist to speed up the process of diagnosis. The ANFIS input consists of features of a X-ray image. It includes contrast, correlation, energy, homogeneity, mean, and variance. The input values enter the first layer of ANFIS, which determines the membership functions belonging to each one. In this step each input has its own group of membership function. The next layer is the rules layer is the second layer which generates the firing strengths for the rules. The outcome of every node is represented by the product of all incoming signals. Then, the third layer, which is the normalization layer, normalizes the firing strengths by computing their sum and dividing each value on it. The next layer (4th layer) takes the normalized values as input as well as the consequence parameters p, q, r. The output is the weighted sum of their inputs added by a constant. The output of this layer is the defuzzification parameters. Finally, the last layer (5th layer–the output layer) computes the average of each rule output giving the final output of the system.
- In [56] a neuro-fuzzy hybridization is considered to take the dual advantages of the neural network and the fuzzy logic. Particularly, a fuzzy logic-based convolutional neural network for Covid-19 diagnosis is used. The input of the system relies on CT scanned images of lungs are considered for classifying the patients into the categories given as Covid-19 and other diseases. The system considers CT scanned images along with other fuzzy parameters and classifies the images into various disease categories using a convolutional neural network. The output is produced after pre-processing, goes to a convolutional neural network for feature extractions. The convolutional neural network works in three phases, namely (i) convolution, (ii) pooling, and (iii) flattened a fully connected neural network. The convolution phase considers the colour image; it needs to be divided into

red, green, and blue colour values. It produces a featured map. In the next phase (pooling), the convolved feature's spatial size is reduced. The aim of this phase is the reduction of the over fitting and the extraction of important representative features from the image. The fully connected neural network now classifies the image into a required category. Also, it adds some symptoms (fever, dry cough, short breath, fatigue, joint pain, sore throat, chills, nausea, diarrhea, etc.). These symptoms are described by trapezoidal and triangular fuzzy sets. Furthermore, in the proposed system there is a fuzzy layer, which, incorporates fuzzy membership functions and fuzzy rules along with the standard fuzzy inference and reasoning. The fuzzy queries and fuzzy training data sets undergo defuzzification with membership functions and various defuzzification techniques encoded into the system within the fuzzy layer. The proposed fuzzy convolutional neural network experimented on 100 CT scanned images of lungs with additional fuzzy symptoms to prove the system's utility.

3.2 Fuzzy Logic-Based Tutoring Systems

Nowadays, advanced technologies and Artificial Intelligence have affected significantly the learning and teaching processes of computer assisted learning systems. They allow tutoring systems to model the student's learning characteristics and needs and provide her/him learning material, instructional advises and techniques that are tailored to them [57–59] Furthermore, they enable tutoring systems to imitate the human's tutor way of thinking and acting during the teaching process [60, 61]. A such advanced technology is fuzzy logic. It deals successfully with the uncertainty, subjectivity and vagueness that characterize the learning process, the student's knowledge level, skills and learning needs determination. There are several scientific studies, which concern the use of fuzzy logic in tutoring systems. The applications of fuzzy logic in educational and computer assisted learning systems concern the following issues of the learning/ teaching processes:

- Student modeling: The fuzzy logic is used to model the student's knowledge level, learning needs and behavior with the aim to deliver her/him the most appropriate learning material and effective instructional method.
- Student's performance prediction: The fuzzy logic is used to predict the student's performance in order to offer better educational services and improve the teaching strategies.
- Student's performance evaluation: The fuzzy logic is used to calculate student's performance considering several subjective and vague assessment factors.
- Adaptive assessment: The fuzzy logic is used for creating assessment tests that are attached to each individual student knowledge level, learning abilities and needs.

3.2.1 Fuzzy-Based Student Models

Student model is a crucial module of each adaptive educational software application and/or computer-based learning system. It allows the system (or software) to identify students' learning characteristics and individual needs, learning characteristics, knowledge level, misconceptions etc. and, according to them, to make instructional decisions that improve the teaching and learning processes. In such a way, learners experience a personalized learning process.

Constructing a student model is a complex process that requires to answer to the following three questions [62]: (i) "What are the characteristics of the user we want to model?, (ii) "How we model them?", (iii) "How we use the user model?". Many researchers, in the literature review, use fuzzy logic for modeling the students' characteristics, answering the 2nd question, and fuzzy rules for taking decisions about the didactic method or content that have to be offered, answering the 3rd question (Table A.3 in Appendix A). For example:

- in [63] fuzzy sets are used to describe each individual student's knowledge level, and fuzzy rules are used for selecting, dynamically, the most appropriate for each individual student learning material and assessment exercises.
- In [64], a novel rule-based fuzzy logic system, which, automatically, model the learning or forgetting process of a student is presented. It uses fuzzy sets to describe the student's knowledge level, a fuzzy logic-based network of concepts to describe the structure and the knowledge dependencies that exist among the several concepts of the learning material, and fuzzy rules to select the most suitable learning path each time for each individual learner.
- In [65] the authors combine fuzzy logic with ontology techniques to model the student's learning behavior. The reason for using fuzzy logic that allows the computational analysis of the student's characteristics and learning behaviors. The aim of this system is to provide an effective learning path to each individual student.
- In [66] the authors use fuzzy logic to manage the uncertainty that characterizes the reasoning process of a student model and to improve the prediction accuracy in student model design.
- In [67], the authors present an adaptive e-learning system for foreign language education, which models the students' knowledge level, preferences and needs, and creates a personalized study plan. The system uses a fuzzy rule-based expert system that takes, for each assessment category, as input (i) the number of correct answers, (ii) the weight of correct answers, (iii) the importance of the category for further studies, (iv) the time spent over answers of the category and gives as output the need of further studies of the given category. For the input and output variables, trapezoidal fuzzy sets are used.
- Another fuzzy logic-based intelligent tutoring system for foreign language learning is presented in [68]. Fuzzy sets are used for the representation of the student's knowledge level on each separate concept of the English and French language. Furthermore, six fuzzy rules are used to determine how the student's

knowledge in English affects her/his knowledge in French, and vice versa. The presented mechanism allows the system to adapt the instructional content to the learner's needs each time.

- In [69], the authors incorporate fuzzy logic in the student model of an Intelligent Tutoring System, which teaches the use of punctuation in Turkish. The presented fuzzy logic-based decision system takes as input the MYCIN Certain Factor, whose value indicates how well is a topic learnt by a particular student, and the number of student's attempts to solve a given problem. The MYCIN Certain Factor input variable is described through seven fuzzy sets and the number of attempts is described through five fuzzy sets. The fuzzy system includes 35 fuzzy rules, that give as output the learning level of the student, which is described by using seven fuzzy sets.
- In [70] an intelligent e-learning system for second language acquisition, which incorporates a fuzzy logic-based inference system, is presented. The fuzzy system uses four trapezoidal fuzzy sets for the representation of the student's knowledge level of each domain concept of the learning material, and 17 fuzzy rules, which represent how the knowledge level of a domain concept in French (or English) is increased or decreased, when the knowledge level of related concepts in English (or French) improves or is getting worse, accordingly. The system decides about the number of exercises of each concept, concerning grammatical issues of tenses and auxiliary verbs of English and French language, that have to be delivered to each individual the learner.
- In [71], the authors present SeisTutor, an Intelligent Tutoring System that adapts the tutoring strategy to the needs of each individual learner. Its adaptation is based on a fuzzy logic-based inference system, which take as inputs the student's knowledge level and learning style and gives as output the learning strategy level. Three fuzzy sets are used for the representation on the knowledge level, 12 fuzzy rules are used for the description of the learning style and 12 fuzzy sets are used for the representation of the learning strategy level. The system uses fuzzy rules of the following format "If knowledge level is X and learning style is Y then learning strategy level is Z".
- In [72], the authors used a hybrid student model, which combines fuzzy logic with Bayesian networks, to identify the students' learning needs and provide adaptive course material. They used a Mamdami Fuzzy Inference System, which takes as input two grades: (i) the "Pretest-Grade", which is the score of the pre-test given to the student to measure her/his prior knowledge in a particular domain, (ii) the "TopicTest-Grade", which is the score of the test given to the student after completing the study of each topic of course material. Three trapezoidal fuzzy sets are used for describing the two input variables. The system's output is "Performance", which is the degree of success in a particular domain topic. The output is described through two trapezoidal fuzzy sets. Nine fuzzy rules of the format "If Pretest-Grade is X and TopicTest-Grade is Y then Performance is Z" are triggered after any change in the value of the test result of a particular topic. Their results determine about the current student's performance level of this topic.

- In [73], the authors developed an Intelligent Tutoring System framework for identifying learning disabilities, like Dyslexia, Dysgraphia and Dyscalculia. The operation of presented framework is based on a fuzzy min–max neural network, which takes as input the learner's score and her/his responding time on a preliminary test and gives as output the category of the learning difficulty, which represent its severity, of and its strength (bound). For the representation of the input variables three triangular fuzzy sets are used. For the description of category, four triangular fuzzy sets are used and for the representation of the bound two triangular fuzzy sets are used. Also, 18 fuzzy rules are used.
- In [74], the authors present an Intelligent Tutoring System, which choose the most appropriate learning environment to deliver the domain knowledge to students with specific learning disabilities. A set of 8 fuzzy rules are used for the evaluation and validation of the selected learning environment. These rules take into consideration the student's attentiveness, excitement and test completion. Their forma is "If Attention status of learner is X and Interest status of learner is Y and Test completion status of learner is Z, then Correctness of provided learning environment is W".
- In [75], an Intelligent Tutoring System for elementary mathematics, which uses fuzzy logic to define the difficulty level of the next math exercise, is presented. It uses a Mamdani Fuzzy Inference System, which has four input linguistic variables: error (the errors that the student made solving math exercises), help (the help that the student asks while solving math exercises), time (the time that the student need to solve math exercises), and emotion (the student's emotion while solving math exercises) and one fuzzy output variable: the difficulty of the next exercise.
- In [76], a fuzzy learning style predictor mechanism is presented. Particularly, a fuzzy decision tree, which takes into consideration student's independent behavior variables during the tutoring conversation is used. Also, a Fuzzy Inference Mechanism that combines grades of membership in all eight dimensions of the four learning styles of the Felder Silverman Learning Styles model and predicts the student's learning style, is used.
- Furthermore, in [77], the authors propose a Neuro Fuzzy Model for the classification of users' interactions within an adaptive e-learning platform, and the identification of the students' learning style. The proposed model is based on Honey and Mumford learning style model, which discerns 4 learning styles: activist, pragmatist, reflector, and theorist. It, also, identifies 20 resource categories, considering the type of learning material resources (i.e. text, multimedia, simulation, concept maps etc.) and the student's preferred action (reading or writing). Three trapezoidal fuzzy sets ("low", "medium", "high") are used for the description of each user's preference for a certain resource category. The output of the Neuro Fuzzy Model detects dynamically the student's learning style and allow the tutoring system to provide to her/him the most appropriate learning material and method.

To sum up, fuzzy logic is an Artificial Intelligence technique, which used to handle the uncertainty that characterizes the detection of students' characteristics, needs and preferences. Fuzzy logic is used to represent the student's knowledge level, cognitive status, learning needs and behaviors, learning styles and preferences. It facilitates the building of the student model, which is crucial for offering adaptive education in a transparent from the learners, manner. The fuzzy logic-based student models contribute to (i) the delivery of appropriate learning and assessment material, (ii) offering personalized and effective learning path, (iii) providing the most appropriate learning environment, (iv) adjusting the tutoring strategy to the learner's needs, (v) improving the prediction accuracy. The fuzzy logic-based student models are, usually, fuzzy-rule based mechanisms. Mamdani Fuzzy Inference System is the most common-used fuzzy-rule based technique that is incorporated in student models. Another preferred technique is Fuzzy Neural Networks.

3.2.2 Fuzzy Logic for Student's Performance Prediction

A crucial factor for design effective courses and study plans is the student's performance prediction. It allows to detect warning signs, shortcomings and inefficiencies regarding the students and the offering courses, and handle with them correcting errors in teaching methods and improving the learning outcomes. Therefore, student's performance prediction provides support to teachers, students, curriculum designers and study consultants. However, student's performance prediction is a complex process, which is depended on several factors, like students' background, learning skills, self-esteem, emotional commitment, motivation, teacher's and course's quality, course's duration e.t.c. All these factors are characterized by subjectivity and vagueness. As a result, fuzzy logic is an ideal way to represent them and model them.

In the literature review, there are several studies about the use of fuzzy logic in the process of student's performance prediction (Table A.4 in Appendix A). For example:

- in [78], a fuzzy logic-based academic performance prediction system is presented. It includes a Takagi–Sugeno fuzzy inference system. The input variables of the fuzzy system are student scores and the number of absences, which are described by fuzzy sets. The system's output is the predicted grade, which is calculated through functions that depended on the input variables. The data analysis provided by the school "Colegio de la Reina" located in Bogotá D.C., Colombia. The aim of the proposed system is to facilitate educational institutions in their understanding of the existing relationships among the different academic degrees and, in turn, to carry out corrective tasks, so that they achieve better academic outcomes, higher educational quality and a better reputation.
- In [79], the authors developed a fuzzy neural network to classify students for predicting academic performance. The neuro fuzzy employees a Takagi–Sugeno fuzzy inference system, which takes three input variables: interest, talent and motivation and generates the student classification. The input variables are described through three fuzzy sets ("low", "medium", high").

- In [80], another neuro fuzzy inference system for prediction of student performance progression is presented. It is based on trends of past performances. The neuro fuzzy model has two inputs: the student's score in the first semester and the student's score in the second semester. Student's scores are represented by using five fussy sets ("Outstanding", "distinction", "average", "below average", and "pass". The model was validated by real time data set comprising of results of mix of students of various higher education institutions under an Indian state university.
- In [81], a Fuzzy Cognitive Map (FCM) was developed for student performance prediction. The FCM includes six factors that concern student performance antecedents: course characteristics, LMS characteristics, student characteristics, student engagement, student support, and institutional factors. Furthermore, three factors, which concern the student's overall performance, are used: score, knowledge construction level, and student Grade Point Average. Through the designed Fuzzy Cognitive Map, the most influential determinants of student performance are identified.
- In [82], a Mamdani FIS is used for the prediction of students' course performance. The system has six linguistic variables as input: daily engagement in studies, participation, English literacy, previous educational achievement, prior knowledge assessment and related course score. These are factors that affect the student course performance, which is the system's output.
- In [83], the authors developed an adaptive neuro fuzzy inference system, which uses multiple parameter sets and a special learning strategy for predicting the future performance of a student. The proposed system used global and local trainings. In addition, it uses premise parameters, which convert crisp values into fuzzy values, and consequence parameters, which compute the output. The number of premise parameters is equal to the number of input variables and the number of consequence parameters is equal to the system's fuzzy rules. The proposed multiple adaptive neuro fuzzy inference system succeeds high accuracy with minimum computational complexity.

Regarding the above, the use of fuzzy logic in students' performance prediction is a relatively new scientific field. According to the search results of Scopus search machine with criteria the words "fuzzy" AND "performance prediction" AND "education" in the article title, abstract and keywords, the first related scientific paper was published in 2015 (Fig. 3.2), but there is an increase in the related published papers last years. Furthermore, the most common-used fuzzy base techniques for the students' performance prediction are the fuzzy neural networks and Takagi–Sugeno FIS.

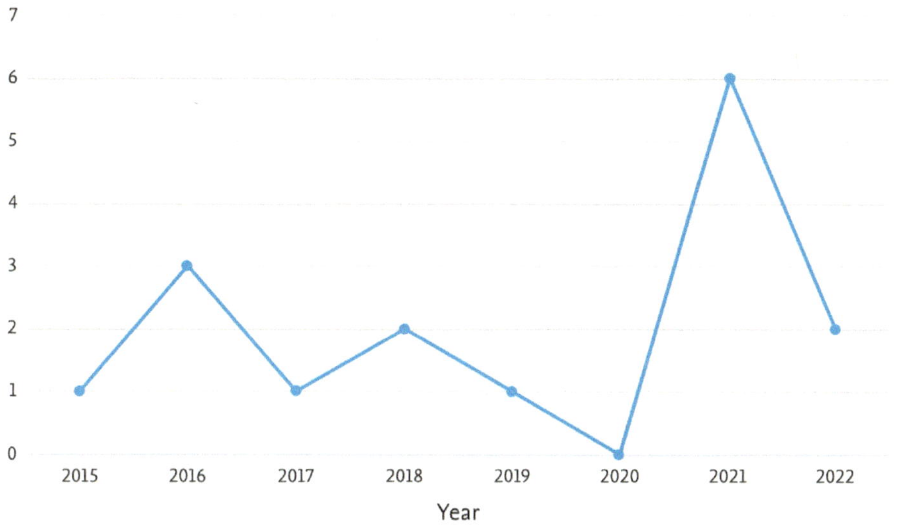

Fig. 3.2 Statistics concerning the published studies in fuzzy logic in students' performance prediction

3.2.3 Fuzzy Logic for Student's Performance Evaluation

Evaluation of student's performance is a key process for identifying the knowledge level, abilities, and misconceptions of students. It is performed, usually, through tests. However, in many cases, student's performance is based not only on the student's score in a test, but on criteria like her/his participation in the learning process, her/his background, her/his assignment achievements, the complexity and difficulty of exercises etc. Therefore, its calculation is a difficult and complex process, which includes human subjectivity and uncertainty. Furthermore, the use of arithmetic values or statistical methods do not explicitly express the level of knowledge acquisition. Fuzzy logic is a technique that allows to handle with this problem. It allows the linguistic expression, which is most ideal for representing knowledge level, skills and abilities. In Table A.5 in Appendix A, several fuzzy logic-based student's evaluation performance systems are represented. Some examples are:

- In [84], a Mamdani FIS for assessing student performance is presented. The system has two inputs: 'Class Record' and 'Exam Performance', and one output 'Student Performance'. The input and output variables of the system are described through triangular membership sets. Three fuzzy sets ('outstanding', 'satisfactory' and 'unsatisfactory') are used for the description of the input 'Class Record', eight fuzzy sets ('Highest', 'Higher', 'High', 'Above Average', 'Average', 'Low', 'Lower' and 'Lowest') are used for the representation of the input 'Exam Performance', and eight fuzzy sets ('Remarkable', 'Excellent', 'Proficient', 'Fair', 'Less Fair', 'Poor', 'Very poor' and 'Fail') are used for the description of the output

'Student Performance'. The system's fuzzy rules are of the format "If (Exam-Performance is X) and (ClassRecord is Y) then (StudentPerformance is Z)". The proposed system uses the centroid approach for defuzzification.

- In [85], the authors introduced a Mamdani FIS to calculate students' performance. The system takes into consideration the importance, complexity, and difficulty of each problem-exercise, which the student solved. These factors are the input values of the system and are described by three triangular and trapezoidal membership fuzzy sets ('low', 'medium', 'high'). The output of the system is the student's comprehensive level of the problem-exercise and is represented by using seven triangular fuzzy sets ('very low', 'low', 'more or less medium', 'medium', 'more or less high', 'high', 'very high'). Twenty-five fuzzy rules of the following format are used: "If importance in X, complexity is Y, difficulty is Z, then comprehensive level is Y". The proposed mechanism calculates students' performance in a way that reflect more accurate the learning situation of students.
- In [86], another Mamdani FIS for students' performance evaluation is presented. The system takes 4 input variables (lecture evaluation, practical evaluation, self-work of the student with the teacher, final evaluation) and gives as output the overall student's evaluation.
- In [87], the authors present a Mamdani FIS, which takes three input variables that are students' marks in three tests received in semester's examinations and gives as output the final student's performance. The system uses five triangular fuzzy membership functions ('very low, 'low', 'average', 'high', (very high') to fuzzify input and other five triangular fuzzy membership functions ('very unsuccessful', 'unsuccessful', 'moderate', 'successful', 'very successful') to fuzzify the output. The initial rule base of the system included 125 fuzzy rules, however a fuzzy rule base compaction (densification) using genetic algorithm is suggested for decreasing the search time and the memory space needed due to the large size of the rule base.
- In [88], the authors performed hybrid clustering to evaluate students' intelligence level in order to create homogeneous groups of students. The proposed approach uses Fuzzy C-Means clustering method and a Takagi–Sugeno FIS, which includes the student's grades in three semesters as input, applies eight fuzzy rules over them and calculates the student's academic performance.
- Similarly, in [89], the authors used Fuzzy C-Means method to classify student academic data. The implementation was done through a web-based mobile application and 4255 records were used. The presented clustering method classified academic students in four categories, according to their overall performance: 'Bad', 'Not Good', 'Very Good', 'Good'.
- In [90], the authors used Fuzzy C-means algorithm to evaluate the progression of the students' behavior by dividing the population into homogeneous groups according to their characteristics and skills. Fuzzy C-means algorithm is used for the automatic generation of the membership functions of the fuzzy sets that describe the students' performance. In addition, a Takagi–Sugeno FIS is used to calculate the overall performance value of a student. The rules of FIS is of the format "If the student belongs to S, then $Z1 = a + bX + 1Y$", where S can be one

of the following linguistic values: very high, high, medium, low, very low; X and Y are the student's grades; and a, b are constants. The proposed system was run with 4,115 university students of Industrial Engineering from a private university in Colombia.

- In [91], the authors designed a Fuzzy Cognitive Map (FCM) to evaluate the learners' performance of practice courses. The concepts of the presented FCM are learning curiosity, desire to learn, preliminary information, learning environment, self-confidence, learning material, course grade 1, course grade 2, course grade 3, practical training, collaborative study, critical task. Their values are fuzzified by three, five or seven membership functions. The input layer of FCM consists of survey results. The second layer consists of exam grades of three courses taught. The output layer consists of critical tasks and concepts measuring the success of the practical courses.
- In [92], a fuzzy logic-based approach for students' performance evaluation and ranking students according to their performance is presented. The students' performance evaluation in the presented study is based on 13 criteria, which are characterized with a different degree of importance. Five triangular fuzzy sets ('very low', 'low', 'medium', 'good', 'very good') are used for the evaluation of the 13 criteria. Five experts evaluate these criteria for each one of the students. Furthermore, the importance of each criterion is described by using three triangular fuzzy sets ('high', 'medium', 'low'). For each student, the general aggregated index is calculated finding the arithmetic mean-based aggregation of the fuzzy numbers that describe her/his evaluation on each criterion, and then, the aggregated values of alternatives are ranked by measuring their distance to the highest term "very good".
- In [93], the authors developed a sample software application for the evaluation of students' performance in engineering education. The students' performance evaluation is based on 5 criteria (presentation, use of materials, teaching ability, technical proficiency, conclusion), which include 24 sub-criteria in total. Each criterion and sub-criterion have weights of significance in evaluation. They are represented by 5 fuzzy sets (excellent (E), good (G), average (A), unsatisfactory (U), poor (P)). The 24 sub-criteria constitute the input variables in a Mamdani FIS. The fuzzy system's output is the 5 main criteria. For the calculation of each output criterion the max-product inference method is used. Also, the center of gravity method is used as defuzzification method. The crisp values of the 5 main criteria, their weights of significance and a multi-criteria calculation method is used, then, to determine the final score of the student's performance.

Therefore, fuzzy logic allows to calculate students' performance by considering subjective and vague criteria and values. It imitates the evaluators' and teachers' way of assessment and expresses the process of performance evaluation in a more explicable way than mathematical and statistical methods. The most common-used fuzzy method for students' performance evaluation is Mamdani FIS. Furthermore, Fuzzy C-Means clustering algorithm is used to create homogeneous groups of students concerning their performance and monitor better the educational process.

3.2.4 Fuzzy Logic for Adaptive Assessment

An important service of computer-based learning systems is the automatic creation of assessment tests. An assessment test is more effective when it includes test items (exercises, quizzes, activities etc.) that are tailored to each individual student's learning needs, abilities and characteristics. Therefore, a computer-based learning system has to incorporate "intelligence", that imitates the human tutor's way of thinking and decision making. An ideal method to perform this "intelligence" is fuzzy logic, since it is able to describe vague and imprecise data, like student's knowledge level, background, learning abilities, characteristics and needs, with linguistic values, and, also, mimics the human way of decision making through a set of linguistic rules [94–97].

There are several computer-based tutoring and learning systems that uses fuzzy logic techniques for selecting the most appropriate test items for each individual student and creating personalized assessment tests (Table A.6 in Appendix A). For example:

- The authors in [98] introduced a new item selection algorithm, based on a multi-criteria decision model, which assesses alternatives according to a set of criteria that were formed by expert's knowledge. The proposed algorithm uses linguistic terms to model the incorporated expert's knowledge. These linguistic terms are defined through fuzzy membership functions. The proposed fuzzy logic-based item selection algorithm enhances the accuracy of assessment and the adaptation of the students' assessment to their competence level.
- In [99] the authors designed 36 fuzzy rules to alter dynamically the complexity level of the assessment test's items. In more details, when a student starts a test, s/he asked to answer to a block of questions, which has a standard complexity. Then, the proposed system considers the student's percentage of right answers, the time that the student needed to answer the questions and the complexity level of the questions, applies the defined fuzzy rules and determines the complexity level of the next block of questions.
- In [100] a fuzzy logic-based algorithm that creates personalized assessment test on the fly is presented. The proposed algorithm was used in a web-based educational application for computer programming instruction. While the student answers to the delivered assessment test, the system decides about the next test item considering the student's misconceptions on previously taught domain concepts and her/his complete or partial knowledge on domain concepts that follow in the teaching process. The system's operation is based on the knowledge dependencies that exist among the different domain concepts of the learning material. The student's knowledge level of each domain is described by using fuzzy sets. Also, the impact of the knowledge of a concept on the knowledge level of a knowledge-related concept is determined by a set of fuzzy rules.
- In [101] the authors combine fuzzy logic and the cognitive theory of Revised Bloom Taxonomy [102] to create effective personalized assessment tests. The following five trapezoidal fuzzy sets are used to describe the student's knowledge

level: "novice", "moderate", "good", "very good", and "expert". The proposed system uses, also, 9 fuzzy rules to define the number and the difficulty level of the test items that have to be included into the created personalized test for each level of the Revised Bloom Taxonomy. The definition of the fuzzy rules was based on the knowledge of experienced tutors.

- In [103] the authors present a fuzzy rule-based system that estimates for each individual student, her/his ability considering the knowledge domain, in order to decide about the next test item that has to be included in the test assessment. The proposed system uses a Mamdani FIS, which accepts as input the item difficulty level, which is described by three trapezoidal fuzzy sets (easy, medium, difficult), the question discrimination, which is described by two trapezoidal fuzzy sets (satisfactory and good), the probability of the student answer the question correctly, which is described by two trapezoidal fuzzy sets (minimum and maximum),and the student's answer, which is described by two trapezoidal fuzzy sets (wrong and right). The proposed system's output is the estimated students' ability, which is described by the following five fuzzy sets: "Very Low", "Low", "Average", "Great", and "Excellent". The membership functions for the fuzzy sets "Very Low" and "Excellent" are trapezoidal. The membership functions of the rest three fuzzy sets are triangular. The output is estimated by applying 24 fuzzy rules, which have the format "If difficulty level is A and question discrimination is B and probability is C and answer is D then ability is E", where A, B, C, D, E are the corresponding defined fuzzy sets. The crisp output is derived using the center of gravity defuzzification method. The value of output determines which will be the next test item.
- In [104], the authors presented a Mamdani FIS for selecting the most appropriate test items from a pool of activities/exercises, for each individual student, in an e-learning system for computer programming. The selection is based on multiple criteria, which constitute the input variables of the presented Mamdani FIS. These variables are: (i) student's knowledge adoption level, (ii) student's prior knowledge, (iii) student's frequency of syntax errors, (iv) student's frequency of logical errors, (v) activity's difficulty level, (vi) activity's relation to syntax errors, (vii) activity's relation to logical errors. The output of the presented fuzzy system is the activity's relevance degree. Input and output variables are represented by trapezoidal fuzzy sets. The system's rule base includes 45 fuzzy rules of the three following formats:

  If student's knowledge adoption level is X and prior knowledge is Y and activity's difficulty level is Z, then the activity's relevance is R. Where X $\in$ {'Poor', 'Moderate', 'High', 'Very High'}, Y $\in$ {'Poor', 'Moderate', 'High'}, Z $\in$ {'Easy', 'Normal', 'Hard'}, R $\in$ {'Low', 'Medium', 'High', 'Very High'}.

  If student's frequency of syntax errors is X and activity's relation to syntax errors is Y, then the activity's relevance is R. Where X and Y $\in$ {'Little', 'Medium', 'High'} and R $\in$ {'Low', 'Medium', 'High', 'Very High'}.

  If student's frequency of logical errors is X and activity's relation to syntax logical is Y, then the activity's relevance is R. Where X and Y $\in$ {'Little', 'Medium', 'High'} and R $\in$ {'Low', 'Medium', 'High', 'Very High'}.

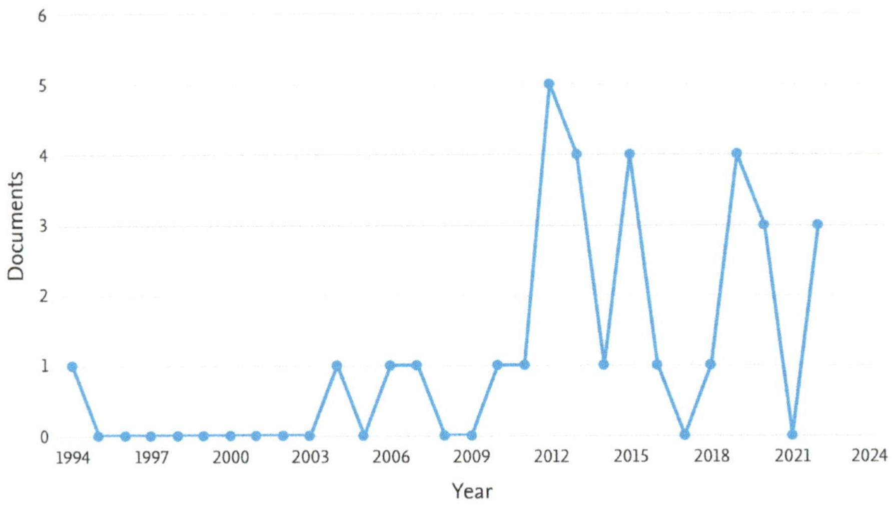

Fig. 3.3 Statistics concerning the published studies in fuzzy logic for adaptive assessment

If more than one rules are applied, then the calculated results from each rule are aggregated. Then, the centroid defuzzification method is used to take the crisp number for the activity's relevance degree. The activities with the highest relevance degree are delivered to the learner.

After a thorough investigation in the related scientific literature, it is obvious that the use of fuzzy logic in the adaptive assessment has not been investigated extensively. According to the search results of Scopus search machine with criteria the words "fuzzy" and "adaptive test" OR "adaptive assessment" OR "personalised test" OR "personalised assessment" in the article title, abstract and keywords, 33 documents were found. An increase in the related published papers is presented in the last 10 years (Fig. 3.3).

3.3 Decision Making Support and Recommended Fuzzy Logic-Based Systems

Decision making is the process of gathering information, defining and assessing alternatives and choosing the best one of them that accomplish a certain goal. It is a reasoning process based on beliefs, preferences, facts and assumptions. Therefore, the decision-making process is characterized by subjectivity and uncertainty. Fuzzy logic allows to model the human reasoning process and convert processes and operations with linguistic terms and imprecise data in mathematically computations. Decision making performs in humans' everyday lives in several fields, like:

- Transportation [105–112].
- Logistics [113–123].
- Sustainability [124–135].
- Supplier selection [136–157].
- Industry-manufacturing [158–175].
- Medical [176–188].
- Risk management [189–196].
- ERP system selection [197–201].
- Investment [202–214].
- Education [215–224].
- Technology-engineering: [158, 225–239].

Fuzzy logic-based decision-making process includes the following steps [1]:

(1) Problem identification: decision-making criteria and options are defined.
(2) Fuzzification: the process of converting variables, which concern criteria and options, into fuzzy values. Fuzzy sets are defined.
(3) Fuzzy interference: definition of the fuzzy rules.
(4) Deffuzification: the process of transferring the fuzzy result (output variable) into a crisp output.
(5) Interpretation and verification: Select the most optimal of the available options regarding the result of the previous step.

There are a variety of fuzzy techniques, methods and operators that are used in decision-making (Table A.7 in Appendix A). For example, Pythagorean fuzzy numbers [240, 241], intuitionistic fuzzy sets [242], q-rung orthopair fuzzy sets [243], which is an extension of intuitionistic fuzzy sets and Pythagorean fuzzy sets [244], Choquet integral, which is an integral created by the French mathematician Gustave Choquet in 1953 [245], as a way to measure the expected utility of an uncertain event [246] are some methods, which used to manage the uncertainty in decision-making process. However, the most common used decision-making methods, which incorporate fuzzy logic, is the multi-criteria decision-making (MCDM) methods. MCDM methods include a well-structured process, which evaluates multiple criteria and finds the most suitable solution-action of a variety of alternative ones. There are several methods of MCDM. Below, the most widely used fuzzy logic-based MCDM methods are presented.

- <u>The hesitant fuzzy linguistic term set (HFLTS)</u>

HFLTS method was introduced in 2012 by Rodriguez, Martinez and Herrera [247]. Since, then the method has known great acceptance. HFLTS supports qualitative decision making, which is used in real-world decision-making problems and situations that confront with uncertainty and subjectivity. HFLTS is a multicriteria decision-making method, in which the experts assess the criteria of each alternative by using linguistic expressions that are based on comparative terms, like 'it is good', 'it is greater than medium', 'it is between 'low' and 'medium', it is lower than 'slow', etc.. The assessment of criteria with linguistic terms is closer to the human way of thinking and contributes to the successful management of making decision problems with uncertainty. An example of a multicriteria decision-making problem is to select the most suitable car to buy. The set of available alternatives is A = {car A, car B, car C, car D), the set of criteria is C = {price, fuel consumption, maintenance cost, safety} and the linguistic term set S = {Very Low (VL), Low (L), Medium (M), High (H), Very High (VH) (Table 3.3).

HFLTS includes three stages:

(1) Transformation: In this stage the experts' linguistic expressions are transformed into Hesitant Linguistic Term Sets by using the transformation function E_{G_H}. According to E_{G_H} the following stand (S is the linguistic term set):

- $E_{G_H}(s_i) = \{s_i\}, where s_i \epsilon S$
- $E_{G_H}(lower than s_i) = \{s_j\}, \forall s_j \leq s_i, s_j \epsilon S, s_i \epsilon S$
- $E_{G_H}(greater than s_i) = \{s_j\}, \forall s_j \geq s_i, s_j \epsilon S, s_i \epsilon S$
- $E_{G_H}(between s_i and s_j) = \{s_k\}, \forall s_i \leq s_k \leq s_j, s_j \epsilon S, s_i \epsilon S, s_k \epsilon S$

  For the example of car selection, the results of the transformation phase are represented in the columns "HFLTS" of the Table 3.4.

(2) Aggregation: In this stage the fusion of the set of criteria for each alternative is performed.For this process, usually, the "min upper" and "max lower" aggregation operators are used. The "min upper" operator selects the worst of the superior values and the "max lower" selects the best of the inferior values. In this way, for each alternative, a linguistic interval, in which the minimum linguistic term is the lower bound and the maximum linguistic value is the upper bound, is defined. For the example of car selection, the results of the aggregation phase are presented in the Table 3.5.

(3) Exploitation: In this stage the defined linguistic intervals are used to choose the best solution of the alternatives. Initially, a method for building a preference relation among the alternatives is used. The proposed method of Rodriguez,

Table 3.3 Alternatives and criteria in the car selection example

	Criteria				
Alternatives		Price	Fuel consumption	Maintenance cost	Safety
	Car A	Between 'H' and 'VH'	Between 'L' and 'M'	Between 'M' and 'H'	Greater that 'H'
	Car B	Between 'M' and 'H'	Between 'M' and 'H'	Greater than 'M'	Greater than 'M'
	Car C	Lower than 'M'	Between 'M' and 'H'	Between 'M' and 'H'	Between 'L' and 'H'
	Car D	H	Lower than M	Between 'L' and 'H'	Greater than 'H'

Table 3.4 The results of the transformation phase in the car selection example

		Criteria											
		Price			Fuel consumption			Maintenance cost			Safety		
		HFLTS	Lower bound	Upper bound	HFLTS	Lower bound	Upper bound	HFLTS	Lower bound	Upper bound	HFLTS	Lower bound	Upper bound
Alternatives	Car A	(H, VH)	H	VH	(L, M)	L	M	(M, H)	M	H	(H, VH)	H	VH
	Car B	(M, H)	M	H	(M, H)	M	H	(M, H, VH)	M	VH	(M, H, VH)	M	VH
	Car C	(VL, L, M)	VL	M	(M, H)	M	H	(M, H)	M	H	(L, M, H)	L	H
	Car D	H	H	H	(VL, L)	VL	L	(L, M, H)	L	H	(H, VH)	H	VH

Table 3.5 The results of the aggregation phase in the car selection example

		Minimum Linguistic term	Maximum Linguistic term	Linguistic intervals
Alternatives	Car A	L	VH	(L, VH)
	Car B	M	VH	(M, VH)
	Car C	VL	H	(VL, H)
	Car D	VL	VH	(VL, VH)

Martinez and Herrera [247] for calculating the preference relation among the alternatives is the following:

$$P_D = \begin{pmatrix} - & p_{12} & \cdots & p1n \\ p_{21} & - & \cdots & p_{2n} \\ \vdots & \vdots & \ddots & \vdots \\ p_{n1} & p_{n2} & \cdots & - \end{pmatrix}, \text{ where } p_{ij} = \frac{\max(0, x_{iR} - x_{jL}) - \max(0, x_{iL} - x_{jR})}{(x_{iR} - x_{iL}) + (x_{jR} - x_{jL})}$$

is the preference degree of the alternative x_i over x_j, $1 \leq i \leq n, 1 \leq j \leq n, i \neq j$, $[x_{iL,}x_{iR}]$ is the linguistic interval of the alternative x_i, $[x_{jL,}x_{jR}]$ is the linguistic interval of the alternative x_j

It is obvious that $p_{ij}+p_{ji}=1$.

For the example of car selection, the results preference relation among the alternatives is the following:

Then, the nondominance degree NDDi for each alternative I is calculates according to the formula $NDD_i = \min\{1 - p_{ji}\}, 1 \leq j \leq nandi \neq j$. The most suitable alternative to choose is the alternative with the maximum NDD. For the example of car selection, the nondominance degrees are:

$$NDD_{CAR\_A} = \min\{(1 - 0.6), (1 - 0.429), (1 - 0.429)\} = 0.4$$
$$NDD_{CAR\_B} = \min\{(1 - 0.4), (1 - 0.333), (1 - 0.333)\} = 0.6$$
$$NDD_{CAR\_C} = \min\{(1 - 0.571), (1 - 0.667), (1 - 0.5)\} = 0.333$$
$$NDD_{CAR\_D} = \min\{(1 - 0.571), (1 - 0.667), (1 - 0.5)\} = 0.333$$

Therefore, the best car to choose is car B.

- Fuzzy Analytic Hierarchy Process (FAHP)

FAHP is a Multicriteria Decision Making method, which combined fuzzy logic with well-known decision-making method Analytic Hierarchy Process, which was introduced by Thomas L. Saaty in the 1970s [248]. FAHP includes the following steps [249]:

(1) Structure the problem: model the problem as a hierarchy defining the decision goal, the decision-making criteria and the alternative solutions.

(2) Construction of a fuzzy pairwise comparison matrix: This is a nxn matrix (n is the number of decision-making criteria), whose cells include a fuzzy set representing the relevant importance the criteria i over the criteria j. Also, the alternatives are pairwise compared against each of the criteria for preference.
(3) Aggregation of the judgements: synthesis of the judgements of multiple experts and calculation of the fuzzy weights of the decision-making criteria. The aggregation of the judgements of multiple experts is performed by applying one of the following fuzzy techniques [250]: the geometric mean method [251, 252], the arithmetic mean method [253, 254], the max–min method [255], a method based on consensus degree [256], the Fuzzy interval geometric mean [257–259], the intuitionistic fuzzy weighted averaging method [260]. The aggregation of the fuzzy weights of the decision-making criteria is performed by applying one of the following methods: the geometric mean method [261], the arithmetic mean method [262], the fuzzy logarithmic least-squares method [263], the lambda-max method [264], the Eigenvector based on index of optimism method [265], fuzzy programming method [266], the fuzzy interval geometric mean method [267, 268], the Atanassov's intuitionistic fuzzy weighted average (IFWA) method [269].
(4) Defuzzification of the fuzzy weights: Convert the fuzzy sets that represent the fuzzy weights of the criteria into crisp values.
(5) Checking for consistency: check if there are judgements between criteria that are in conflict (for example, lets criterion X is more important than criterion Y, according to logical inference computations, but experts' opinion is than criterion X is less important than criterion Y). In case of no consistency, the experts have to judge again the decision-making criteria.
(6) Come to a final decision based on the results of the above process.

Steps 1 and 6 are absolutely similar in AHP and FAHP. Steps 2, 3 and 5 are included in both AHP and FAHP, but different techniques are applied in the calculations. Step 4 is performed only in FAHP.

- Fuzzy Technique for Order of Preference by Similarity to Ideal Solution (Fuzzy TOPSIS)

Fuzzy TOPSIS is a Multicriteria Decision Making method, which integrates fuzzy logic in TOPSIS method, which was introduced by Ching-Lai Hwang and Yoon in 1981 [270] and is based on the idea that the optimum alternative has the least geometric distance from the positive ideal solution and the longest geometric distance from the negative ideal solution. An example of a multicriteria decision-making problem, that can be solved by using fuzzy TOPSIS, is to choose the better student of a class. The set of available alternatives is A = {Maria, George, Mike, Kate), the set of criteria is C = {knowledge, participation, cooperation, comprehension, time to solve exercises} and the linguistic term set S = {Very Low (VL), Low (L), Average (Av), High (H), Very High (VH)}. The membership functions of the defined fuzzy sets are depicted in Table 3.6.

The steps of fuzzy TOPSIS follow [271]:

Table 3.6 The fuzzy sets and their membership functions in the best student selection example

Fuzzy sets—linguistic value	Membership function partition
Very low (VL)	(1, 1, 3)
Low (L)	(1, 3, 5)
Average (Av)	(3, 5, 7)
High (H)	(5, 7, 10)
Very High (VH)	(7, 10, 10)

(1) Create an evaluation matrix X[mxn], where m is the number of alternatives and n is the number of criteria. This matrix includes the experts' linguistic ratings for each alternative and for each criterion xij. Tables 3.7 and 3.8 depict the evaluation matrix for the example of the students' assessment. In this example, the experts were three. Also, experts define the fuzzy weights for criteria. Table 3.9 depicts the criteria weights.

(2) Aggregation of experts' fuzzy ratings and criteria's fuzzy weights. The aggregation of the experts' fuzzy rating of alternative i for criterion j is performed by the formula:$a_{ij} - \min_k \left\{ a_{ij}^k \right\},\ b_{ij} = \frac{1}{K} \sum_{k=1}^{K} b_{ij}^k,\ c_{ij} = \max_k \left\{ c_{ij}^k \right\}$

Therefore, the aggregated value is (a_{ij}, b_{ij}, c_{ij}). The aggregated evaluation matrix for the example of students' assessment is depicted in Table 3.10.

The aggregation of the fuzzy weights of criterion j is performed by the formula:$w_{j1} - \min_k \left\{ w_{j1}^k \right\},\ w_{j2} = \frac{1}{K} \sum_{k=1}^{K} w_{j2}^k,\ w_{j3} = \min_k \left\{ w_{j3}^k \right\}$

Therefore, the aggregated weight is (w_{j1}, w_{j2}, w_{j3}). The aggregated evaluation matrix for the example of students' assessment is depicted in Table 3.11.

(1) Create the normalized fuzzy decision matrix R = [rij], according to the following formulas:

- for benefit criteria (criteria which impact positively on the solution)$r_{ij} = \left(\frac{a_{ij}}{c_j^*} \frac{b_{ij}}{c_j^*}, \frac{c_{ij}}{c_j^*} \right),\ c_j^* = \max\left\{ c_{ij} \right\}$

Table 3.7 The evaluation matrix for the best student selection example—linguistic values

		Criteria				
		Knowledge	Participation	Cooperation	Comprehension	Time to solve exercises
Alternatives	Maria	H, H, VH	H, A, H	A, H, H	H, VH, VH	A, A, L
	George	A, A, H	H, VH, VH	H, H, VH	A, H, H	VL, L, L
	Mike	VL, L, L	L, A, A	L, A, A	L, L, A	H, VH, H
	Kate	L, A, A	A, A, H	H, VH, VH	A, H, H	A, A, H

Table 3.8 The evaluation matrix for the best student selection example – Partition of the fuzzy sets

		Criteria				
		Knowledge	Participation	Cooperation	Comprehension	Time to solve exercises
Alternatives	Maria	(5, 7, 10), (5, 7, 10), (7, 10, 10)	(5, 7, 10), (3, 5, 7), (5, 7, 10)	(3, 5, 7), (5, 7, 10), (5, 7, 10)	(5, 7, 10), (7, 10, 10), (7, 10, 10)	(3, 5, 7), (3, 5, 7), (1, 3, 5)
	George	(3, 5, 7), (3, 5, 7), (5, 7, 10)	(5, 7, 10), (7, 10, 10), (7, 10, 10)	(5, 7, 10), (5, 7, 10), (7, 10, 10)	(3, 5, 7), (5, 7, 10), (5, 7, 10)	(1, 1, 3), (1, 3, 5), (1, 3, 5)
	Mike	(1, 1, 3), (1, 3, 5), (1, 3, 5)	(1, 3, 5), (3, 5, 7), (3, 5, 7)	(1, 3, 5), (3, 5, 7), (3, 5, 7)	(1, 3, 5), (1, 3, 5), (3, 5. 7)	(5, 7, 10), (7, 10, 10), (5, 7, 10)
	Kate	(1, 3, 5), (3, 5, 7), (3, 5, 7)	(3, 5, 7), (3, 5, 7), (5, 7, 10)	(5, 7, 10), (7, 10, 10), (7, 10, 10)	(3, 5, 7), (5, 7, 10), (5, 7, 10)	(3, 5, 7), (3, 5, 7), (5, 7, 10)

Table 3.9 The criteria weights for the best student selection example

		Expert 1	Expert 2	Expert 3
Knowledge	Linguistic value	VH	VH	H
	Fuzzy interval	(7, 10, 10)	(7, 10, 10)	(5, 7, 10)
Participation	Linguistic value	VH	A	H
	Fuzzy interval	(7, 10, 10)	(3, 5, 7)	(5, 7, 10)
Cooperation	Linguistic value	H	A	L
	Fuzzy interval	(5, 7, 10)	(3, 5, 7)	(1, 3, 5)
Comprehension	Linguistic value	H	VH	H
	Fuzzy interval	(5, 7, 10)	(7, 10, 10)	(5, 7, 10)
Time to solve exercises	Linguistic value	L	A	VL
	Fuzzy interval	(1, 3, 5)	(3, 5, 7)	(1, 1, 3)

– for cost criteria (criteria which impact negatively on the solution):$r_{ij} = \left(\frac{a_j^-}{c_{ij}}\frac{a_j^-}{b_{ij}}, \frac{a_j^-}{a_{ij}}\right)$, where $a_j^- = \min\{a_{ij}\}$

The normalized fuzzy decision matrix for the example of students' assessment is depicted in Table 3.12 ("time to solve exercises" is a cost criterion, the rest criteria are benefit criteria).

(2) Create the weighted normalized fuzzy decision matrix. Each cell v_{ij} equals to $r_{ij}*w_j$. The weighted normalized fuzzy decision matrix for the example of students' assessment is depicted in Table 3.13.

Table 3.10 The aggregated evaluation matrix for the best student selection example

		Criteria				
		Knowledge	Participation	Cooperation	Comprehension	Time to solve exercises
Alternatives	Maria	(5, 8, 10)	(5, 6.33, 10)	(3, 6.33, 10)	(5, 9, 10)	(1, 4.33, 7)
	George	(3, 5.667, 10)	(5, 9, 10)	(5, 8, 10)	(3, 6.33, 10)	(1, 2.33, 5)
	Mike	(1, 2.33, 5)	(1, 4.33, 7)	(1, 4.33, 7)	(1, 3.667, 7)	(5, 8, 10)
	Kate	(1, 4.33, 7)	(3, 5.667, 10)	(5, 9, 10)	(3, 6.33, 10)	(3, 5.667, 10)

Table 3.11 The aggregated weights for the best student selection example

	Fuzzy weight
Knowledge	(7, 9, 10)
Participation	(3, 7.33, 10)
Cooperation	(1, 5, 10)
Comprehension	(5, 8, 10)
Time to solve exercises	(1, 3, 7)

Table 3.12 The normalized fuzzy decision matrix for the best student selection example

		criteria				
		Knowledge	Participation	Cooperation	Comprehension	Time to solve exercises
Alternatives	Maria	(0.5, 0.8, 1)	(0.5, 0.633, 1)	(0.3, 0.633, 1)	(0.5, 0.9, 1)	(0.143, 0.231, 1)
	George	(0.3, 0.5667, 1)	(0.5, 0.9, 1)	(0.5, 0.8, 1)	(0.3, 0.633, 1)	(0.2, 0.429, 1)
	Mike	(0.1, 0.233, 0.5)	(0.1, 0.433, 0.7)	(0.1, 0.433, 0.7)	(0.1, 0.3667, 0.7)	(0.1, 0.125, 0.2)
	Kate	(0.1, 0.433, 0.7)	(0.3, 0.5667, 1)	(0.5, 0.9, 1)	(0.3, 0.633, 1)	(0.1, 0.177, 0.33)

(3) Compute the Fuzzy Positive Ideal Solution (FPIS) and Fuzzy Negative Ideal Solution (FNIS), according to the following formulas:

$$A^* = \left(\tilde{v}_1^*, \tilde{v}_2^*, \ldots \tilde{v}_n^*\right), \quad \text{where } \tilde{v}_j^* = \max_i \{v_{ij3}\}$$

$$A^- = \left(\tilde{v}_1^-, \tilde{v}_2^-, \ldots \tilde{v}_n^-\right), \quad \text{where } \tilde{v}_j^- = \max_i \{v_{ij1}\}$$

Table 3.13 The weighted normalized fuzzy decision matrix for the best student selection example

		Criteria				
		Knowledge	Participation	Cooperation	Comprehension	Time to solve exercises
Alternatives	Maria	(3.5, 7.2, 10)	(1.5, 4.64, 10)	(0.3, 3.165, 10)	(2.5, 7.2, 10)	(0.143, 0.693, 7)
	George	(2.1, 5.1, 10)	(1.5, 6.597, 10)	(0.5, 4, 10)	(1.5, 5.064, 10)	(0.2, 1.287, 7)
	Mike	(0.7, 2.097, 5)	(0.3, 3.174, 7)	(0.1, 2.165, 7)	(0.5, 2.9336, 7)	(0.1, 0.375, 1.4)
	Kate	(0.7, 3.897, 7)	(0.9, 4.154, 10)	(0.5, 4.5, 10)	(1.5, 5.064, 10)	(0.1, 0.531, 2.31)

For the example of students' assessment, FPIS and FNIS are presented in Table 3.14.

(4) Compute the distance from each alternative to the FPIS and to the FNIS, according to the following formulas, respectively.

Table 3.14 The Fuzzy Positive Ideal Solution (FPIS) and Fuzzy Negative Ideal Solution (FNIS) for the best student selection example

		Criteria				
		Knowledge	Participation	Cooperation	Comprehension	Time to solve exercises
Alternatives	Maria	(3.5, 7.2, 10)	(1.5, 4.64, 10)	(0.3, 3.165, 10)	(2.5, 7.2, 10)	(0.143, 0.693, 7)
	George	(2.1, 5.1, 10)	(1.5, 6.597, 10)	(0.5, 4, 10)	(1.5, 5.064, 10)	(0.2, 1.287, 7)
	Mike	(0.7, 2.097, 5)	(0.3, 3.174, 7)	(0.1, 2.165, 7)	(0.5, 2.9336, 7)	(0.1, 0.375, 1.4)
	Kate	(0.7, 3.897, 7)	(0.9, 4.154, 10)	(0.5, 4.5, 10)	(1.5, 5.064, 10)	(0.1, 0.531, 2.31)
	FPIS (A^*)	(3.5, 7.2, 10)	(1.5, 6.597, 10)	(0.5, 4.5, 10)	(2.5, 7.2, 10)	(0.2, 1.287, 7)
	FNIS (A^-)	(0.7, 2.097, 5)	(0.3, 3.174, 7)	(0.1, 2.165, 7)	(0.5, 2.9336, 7)	(0.1, 0.375, 1.4)

$$d_i^* = \sum_{j=1}^{n} d\left(\tilde{v}_{ij}, \tilde{v}_j^*\right),\quad d_j^- = \sum_{j=1}^{n} d\left(\tilde{v}_{ij}, \tilde{v}_j^-\right)$$

For the example of students' assessment, the distance of each alternative to FPIS is presented in Table 3.15 and the distance of each alternative to FNIS is presented in Table 3.16.

(5) Compute the closeness coefficient CCi for each alternative, according to the formula $CC_i = \frac{d_i^-}{d_i^- + d_i^*}$. The CCi for each alternative of the example of students' assessment is presented in Table 3.17.

(6) Rank the alternatives. The alternative with highest closeness coefficient represents the best alternative. Therefore, in the example of students' assessment, the best student is Maria and George follows.

Table 3.15 The distance of each alternative to FPIS for the best student selection example

		Criteria					
		knowledge	Participation	Cooperation	Comprehension	Time to solve exercises	d_i^*
Alternatives	Maria	0	1.13	0.779	0	0.345	2.254
	George	1.457	0	0.289	1.362	0	3.108
	Mike	4.43	2.718	2.207	3.225	3.276	15.856
	Kate	3.041	1.452	0	1.362	2.743	8.598

Table 3.16 The distance of each alternative to FNIS for the best student selection example

		Criteria					
		knowledge	Participation	Cooperation	Comprehension	Time to solve exercises	d_i^-
Alternatives	Maria	4.43	3.087	1.829	3.225	3.239	15.81
	George	3.463	3.566	2.044	2.201	3.276	14.55
	Mike	0	0	0	0	0	0
	Kate	1.554	2.962	2.207	2.201	0.533	9.2

Table 3.17 The CCi for each alternative for the best student selection example

Alternatives	Closeness coefficient
Maria	0.8752
George	0.824
Mike	0
Kate	0.517

3.4 Fuzzy Logic-Based Natural Language Processing Systems

Natural language processing (NLP) systems belong to artificial intelligence field. They combine methods and theories of three disciplinary fields: computer science, linguistics science and artificial intelligence (Fig. 3.4). They aim to make computers able to understand the human language and communicate with them. Particularly, a NPL system is able to recognize speech or read documents, understand their content and extract important information from them. Some examples of NPL systems are: smart assistants, language translators, automated customer service systems, email filters, text analytics tools, predicting text applications, chatbot applications etc.

Analyzing and understanding the content of a speech or text is a complex process, which is characterized by uncertainty and vagueness. Fuzzy logic, which is a computational intelligence technique that allows a computer to compute with words and imitate the human way of thinking, is ideal to manage the uncertainty and vagueness of natural language. For example, according to Gupta, Jain and Joshi [272] fuzzy logic helps in understanding the context of a text in a better way by modeling vague words, like some, little, most etc. The authors illustrate representative examples of Google Search Engine, Google Translator and MIT Start.

In the recent literature review, there are several applications of fuzzy logic in NPL systems (Table A.8 in Appendix A). Particularly, they include chatbot applications for medical diagnosis, geolocation applications, customer service applications, speech recognition systems, text summarization, text analysis applications etc. (Fig. 3.5) Especially, fuzzy logic is used in sentiment analysis, which belongs to NLP. More specifically, sentiment analysis is referred to the extraction and identification of emotions, perceptions and opinions about a product, service, event etc. from texts.

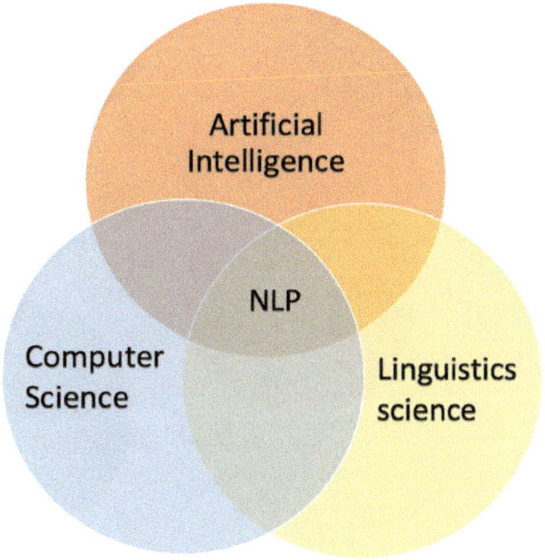

Fig. 3.4 Fields of Natural Language Processing (NPL)

In sentiment analysis, fuzzy logic is, usually, used for the classification of emotions or opinions into positive, negative and neutral.

Below, some systems of the recent literature, that use fuzzy logic in the NPL are presented.

- In [273], the authors developed a fuzzy logic-based chatbot service for medical diagnosis, which was incorporated in the Covenant University Doctor (CUDoctor) telehealth system. The aim of the developed service is to ask the user her/his symptoms, recognize the user's answers, and send a notification message to the user about the diagnosis, which was performed by CUDoctor. It incoprorated a fuzzy inference module (Mamdani Fuzzy Inference system and weighted fuzzy rules), which takes as input the user's answers, contributes to keyword extraction and matching, tracks her/his symptoms and decides about which is the most relevant question that has to be posed next to the user.

Fig. 3.5 Applications of fuzzy logic-based Natural Language Processing Systems

- In [274], the authors used fuzzy logic in Natural language processing to interpret some vague questions concerning geolocation, like "I would like to be notified when a taxi is close to the square". Particularly, firstly a lexicon is constructed. Each lexicon input is attached a tag that indicated its grammatical category (noun, verb, adverb, etc.). Then, synonyms of the lexicon words are defined concerning their semantic distance. For each imprecise concept a model inspired by the 2-tuple fuzzy linguistic representation model by Herrera and Mart´ınez [275] is used.
- In [276], the authors developed a system for automatic detection of plagiarism in academia and scientific research documents, which uses fuzzy logic for identifying semantic similarities among compared documents. Initially, tokenization, tagging, pruning and lemmatization of words. Then, an improved fuzzy-semantic similarity measure method, which uses different ranges of similarity scores, is applied. Comparisons are done between words of same classes only.
- In [277], the authors developed a fuzzy logic-based NLP system, which aims to predict the words failed to be recognized by a speech recognition system. The presented fuzzy logic-based system identifies if a recognized word is semantically appropriate with respect to the whole recognized sentence and predicts possible missing words by applying a set of fuzzy rules.
- In [278], the authors, presented a system that identifies the customers' emotions of a restaurant analyzing their reviews on social media. The authors used fuzzy sets with triangular membership functions and fuzzy rules to encode data in Long Short-Term Memory method.
- In [279], the authors presented a chatbot, which asks users to enter in text details concerning themselves, historical medical data and symptoms and predicts their possible disease and the appropriate medicine. The operation of the presented chatbot is based on fuzzy support vector machine algorithm.
- In [280], the authors developed a system that facilitates the recruitment process of an organization. In more details, the proposed system matches, automatically, CVs to the descriptions of offered job positions and, also, provides an estimation of the overall CV/job offer expertise level. This is performed through NLP, an ontology-based approach, and a fuzzy logic-based model. Regarding the embedded fuzzy logic-based model, it is used to weight the relationship between technology skills and domains. Particularly, a Mamdani FIS is used. It takes as input tree parameters, concerning the matching process, and applying a set of fuzzy rules, gives as output the expertise level weight of the job offer according to the activity verbs in CV.
- In [281], the authors developed a medical web-based application, which is called MEDSHARE, to integrate patients' healthcare records from heterogeneous sources and offer diagnose services. The system informs the users about the possible disease through mobile or web-based notification. The system's operation includes data collection, data process, symptoms extraction, and keyword mapping. Then, a Mamdani FIS with multiple fuzzy rules is used to identify the diseases that a user-patient is prone to.

- In [282], the authors proposed a Mamdani FIS for summarizing texts in English and German and in two Indian languages: Marathi and Hindi. The fuzzy inference system takes as input eight parameters: title (the degree of similarity between title and every other sentence in the document), sentence length (the length of each sentence divided by the length of the maximum sentence of the document), sentence position (the rank of the sentence divided by the number of document's sentences), sentence numerical data (statistic information about each sentence), thematic words (rank each sentence according to the occurrence in it of the most frequent displayed words in the document), sentence to sentence similarity, term weight (the sum of frequency of each word displayed in a sentence divided by the maximum summation of all sentences of the document), proper nouns (total number of proper nouns in a sentence divided by the sentence length). The output of the fuzzy inference system is the importance of each sentence of the document. The higher important sentences are selected for forming the document's summary.
- In [283], the authors use a Mamdani Fuzzy Inference System to select the most important sentences of a text, which will used to generate the text's summary. The fuzzy inference system takes as input six features, which characterize each sentence of the overall text. These features are: title feature (the sentence's words that appear, also, in the document's title), sentence length, term weight (the frequency of term occurrences), sentence position, proper noun (the ration of sentence's proper nouns to the sentence's length), numerical data. They take a value between 0 and 1 and are described through fuzzy sets with triangular membership function. The system applies over these features, fuzzy IF–THEN rules and give as output the importance of each sentence. The format of the system's fuzzy rules is "If 'title feature' is Very High and 'sentence length' is High and 'term weight' is Very High and 'sentence position' is High and 'proper noun' is Very High and 'numerical data' is Very High, then 'sentence' is Important. The system's output is described by three fuzzy sets: 'Unimportant', 'Average', 'Important'. For the defuzzification of the output, the centroid method is used.
- In [284], the authors present a system, which extracts users' emotions from textual data in social media. The authors use fuzzy logic to deal with the ambiguity of emotions and the fact that a sentence in a text can reveal multiple emotions at the same time. Particularly, they use Natural Language Processing, which incorporates Fuzzy Neural Networks and Deep Recurrent Neural Networks. The presented Fuzzy Neural Network includes four layers. The first layer, which is the input layer, takes as input the feature-representation vector of the semantic and structural feature extraction phase of the proposed method. The second layer of represents the degree of belonging of different features to different emotional classes in datasets. The third layer corresponds to the fuzzy rules. The fourth layer includes the output fuzzy variables. In this layer, the number of neuros is equal to the number of the emotional classes in datasets. The last layer includes the defuzzification phase.
- In [285], the authors present a fuzzy logic-based model, which identifies positive, neutral or negative perceptions from text and emoji comments in social media.

They use fuzzy sets and fuzzy rules to represent and analyze the consumers' tweets for marketing purposes.

- In [286], the authors present a fuzzy-based NPL approach for sentimental analysis from social media posts. Particularly, the general proposed sentiment analysis approach includes three phases: text pre-processing, use of sentiment lexicon and fuzzy rule system for sentiment polarity classification. In the third phase, a Mamdani Fuzzy Inference System is used. The fuzzy system takes as input two fuzzy variables: the post's 'positive score' and the post's 'negative score'. For the representation of them, three fuzzy sets ('Low', 'Medium', 'High') with triangular membership functions. Nine fuzzy rules are applied over these fuzzy variables to define the sentiment, which is the system's output. The output is described by three fuzzy sets ('negative', 'neutral', 'positive'). The format of the applied fuzzy rules is the following: "If 'positive score' is X and 'negative score' is Y, then 'sentiment' is Z".
- In [287], the authors propose the use of Sugeno-Tsukamoto fuzzy inference system for sentiment analysis. They choose this kind of fuzzy inference system because it is applied usually to classification problems. Through, this model they reduce the computational complexity and increase the interpretability of sentiment analysis. For the representation of the system's input, trapezoid fuzzy membership functions are proposed.
- In [288], the authors present a sentimental fuzzy logic-based analysis method for extracting emotions and opinions from audio reviews. The system, initially, converts the audio file, which has been submitted by the user, into text. Then, text processing and document classification follow. Then, fuzzy logic is used to classify the review as positive, negative or neutral. The proposed fuzzy logic-based approach includes linguistic representation for each polarity and concerns lemmatized word and their synonyms.

3.5 Fuzzy Logic in Automotive Systems

An automotive system is any vehicle's system, which includes a number of electronic components that work together to fulfil a specific task that contributes to safe, comfort and fuel/energy efficient driving. Automotive systems are:

- The brake system.
- The fuel system.
- The transmission system.
- The suspension system.
- The engine system.
- The cooling system.
- The emission system.

Factors that affect the safe and fuel-efficient driving are confront with uncertainty and inaccuracy. For example, in each circumstance, the driver has to decide "how much have I to accelerate?", "what is the appropriate speed for less fuel consumption?", "what is the safest distance from the vehicle in front?", "when have I to brake to avoid an obstacle?". There is not a clear and objective way to manage these situations. These decisions cannot be based on accurate mathematical models. They are based on subjectivity and driving experience. As a consequence, fuzzy logic is ideal to be used as reasoning and computational method for designing intelligent driving controllers.

The use of fuzzy logic and the development of fuzzy systems in areas related to vehicles and automotive begins from 1980s and increased from early 1990s [289]. Recent years, there is an increased interest in applying fuzzy logic techniques for design intelligent controllers for automotive systems. Examples of fuzzy logic-based intelligent driving controllers include anti-collision controller, speed and acceleration controller, cruise controller, brake assist system controller, collision controller, suspension system controller, fuel or energy consumption controller, driving behavior controller etc.

After a search in recent scientific bibliography, we decide to separate the areas of automotive systems, in which fuzzy logic is applied in the following four categories:

(i) **assisted driving**: it includes the automation of systems that help a driver to drive safety and efficient.
(ii) **driver modelling**: it includes the detection and identification of driving behavior.
(iii) **energy/fuel efficiency**: it includes the intelligent management of fuel or energy consumption and carbon emission.
(iv) **ride comfort**: it includes the automation of systems that make the driving more easy and pleasant.

Below, several studies, concerning the use of fuzzy logic in automotive systems, are described. The studies are presented per previously defined automotive area category. An overall presentation of the studies is depicted in Table A.9 in Appendix A.

Assisted driving

- In [290], the authors proposed an approach, which uses fuzzy rough sets, for fault detection and recovery in a fault-tolerant system. The fuzzy-base control is performed to detect the exact location and nature of fault in terms of states in the automotive system.
- In [291], a fuzzy logic-based anti-collision control strategy is presented. Particularly, a Mamdani Fuzzy Inference System, which takes as input the difference between the expected distance and the actual distance and the relative speed of the two vehicles and gives as output the amount of change in the throttle control, is presented.
- In [292], a fuzzy logic-based approach for vehicle speed estimation is presented. Particularly, the proposed system is based on Mamdani Fuzzy Inference System. It takes as input the driving condition, the difference between the tire-ground contact patch and the vehicle velocity, and the maximum deviation between the faster and

the slower tire-ground contact patch velocity. The system concerns the following five driving conditions: strong braking, braking, coasting, acceleration, and strong acceleration. The system's input variables are expressed through linguistic values, which are represented by trapezoidal membership functions. The system uses 32 fuzzy rules to evaluate the level of credibility of the measured velocity of each wheel and the level of the longitudinal acceleration. The output variables are used to estimate the vehicle's speed. The fuzzification of the system's output is performed using the centroid method.

- In [293], the authors embed a Takagi–Sugeuo fuzzy system in a sliding mode wheel slip ratio controller for intelligent anti-lock controlling of vehicle wheels.
- In [294], a Takagi–Sugeno fuzzy inference system is used for intelligent control of the braking system. Particularly, the proposed system takes as input the velocity, the deceleration, the measured distance to the current traffic stimulus, traffic light, the velocity of the car in front, braking distance – distance rate, and gives as outputs the current velocity and the distance change.
- In [295], a fuzzy rule-based control algorithm for anti-lock braking system is presented. The proposed system takes as input the automobile adhesion coefficient and the wheel slip ratio and applying fuzzy rules, which embeds the knowledge and experience of engineers, calculates the vehicle's braking force.
- In [296], the authors propose a Mamdani Fuzzy Inference System, which estimates the risk level of collision with a pedestrian. The proposed system takes as input the pedestrian's position, distance, movement direction, and movement magnitude and gives as output the risk level. Input and output fuzzy variables are modeled through triangular and trapezoidal membership functions. The values of the input variables are extracted by images captured by an on-dash camera in the car using computer vision technologies.
- In [297], the authors use a fuzzy 3-layer neural network to estimate the desired acceleration of a vehicle, when it faces obstacles. The proposed fuzzy neural network has two input variables: the relative velocity and distance of the vehicle against the obstacle and gives as output the target acceleration. The neural networks' input and output variables are described through triangular and trapezoidal membership functions.

Driver modelling

In [298], the authors present an intelligent method for calculating the risk of driving behavior, which is based on fuzzy logic and Back Propagation Neural Network. The system checks the driver's face features, expressions and direction to detect the risk. The classification of risk alert is performed through a Mamdani Fuzzy Inference System, which takes as input the driver's facial characteristics and expressions, that are detected in a previous phase, and her/his blinking speed, and gives as output the risk level. The risk level takes a linguistic value from the set {'Very Low', 'Low', 'Moderate', 'High', 'Very High'}. Some examples of the applied fuzzy rules are the following: "If normal blinking, then the risk level is low", "if there is lower blinking and distraction less than 1 s, then the risk level is moderate", "if there is drowsiness and the distraction lasts more than 2 s then the risk level is very high". According to the risk level, an alarm is generated.

Energy/fuel efficiency

- In [299], a Mamdani Fuzzy Inference System, which estimates the range of an electric vehicle, is presented. The fuzzy system takes as input the battery charge status and the power consumed and gives as output how far the electric vehicle can travel.
- In [300], a fuzzy logic-based intelligent energy management system for hybrid electric autonomous vehicles is presented. Particularly, an Interval Type-2 fuzzy logic system is used for managing the energy consumption.
- In [301], the authors developed a fuzzy logic-based energy monitoring system for hybrid electrical vehicles. Particularly, a Sugeno fuzzy inference model in combination with particle swarm optimization methos are used. The input variables of the fuzzy system are the required propulsion power and the state of charge of the battery and are described through trapezoidal membership functions. The system applies 15 rules over the input variables and give as output the fuel cell reference power.
- In [302], two fuzzy logic controllers for energy management in electrical vehicles are presented. The controllers are based on Mamdani Fuzzy Inference System. The first one take as input the aggregate state of charge of the vehicle and the nodal voltage of vehicle's charging station and gives as output the aggregate energy of vehicle's charging station. The output of the first fuzzy controller along with the nodal voltage secondary winding distribution transformer are passed as inputs to the 2nd fuzzy controller, which gives as output the vehicle's charging or discharging power. Therefore, the two fuzzy systems cooperate to decide the proper energy flow between the electric vehicle and the grid.
- In [303], a fuzzy logic-based controller for idle speed controlling is presented. The proposed approach aims to keep stable the vehicle's speed managing efficiently the fuel consumption. The system takes as input the idle speed deviation and the rate of change of the speed deviation and applying fuzzy rules attempts to control the vehicle's idle speed. Gaussian and triangular membership functions describe the fuzzy input and output variables of the system.
- In [304], an intelligent controller for determining the vehicle's acceleration and its safe distance from the lead vehicle is presented. The proposed system is based on an Adaptive Neuro Fuzzy Inference System with five layers, whose rules are based on a Tankagi–Sugeno model.
- In [305] the authors developed an intelligent fuzzy logic-based cruise controller. The proposed system is based on Mamdani Fuzzy Inference System. It takes as input two linguistic variables: error and delta error, which are described through triangular and trapezoidal membership functions. The system gives as output the change in Pulse Width Modulation duty cycle of the vehicle's motor. The output is generated by applying nine fuzzy rules.

Ride comfort

- In [306], a fuzzy control strategy for semi-active suspensive to effectively improve the vibration reduction effect of the automobile sus- pension system is presented. The fuzzy control is based on a Mamdani Fuzzy Inference System.
- In [307], introduced intelligence in driving through a Takagi–Sugeno fuzzy model and an adaptive neuro-fuzzy predictor-based control approach. The Takagi–Sugeno fuzzy model takes as input the position, velocity and acceleration of the preceding vehicle and predicts its state sequence. The proposed adaptive neuro-fuzzy approach controls the driving of the following vehicle. In such way, safe keeping of the desired distance between the two vehicles, stability, and efficient fuel consumption are achieved.
- In [308], a 5-layer Adaptive Neuro Fuzzy Inference System (ANFIS) for controlling the vibration of a vehicle suspension system. The input of the proposed ANFIS is described through five bell-shaped membership functions. The system's output is the suspension performance and is described through a linear type of membership function.

Regarding the above, the most application of fuzzy logic in automotive systems concern the assisted driving and fuel/energy efficiency. Furthermore, Mamdani Fuzzy Inference System is the most common-used fuzzy technique and Takagi–Sugeno Fuzzy Inference System is the second common-used fuzzy technique. Also, most automotive systems regard the control of vehicle's speed and acceleration. In addition, in more studies no other technique is used in combination with fuzzy logic to develop an intelligent driving control system.

3.6 Scheduling, Network Traffic and Security Fuzzy Logic-Based Software Systems

3.6.1 Fuzzy Logic in Task Scheduling

In computer science, task scheduling is called the process of selecting each time the task, which has to utilize the computer's central processing unit and for which the appropriate for its execution computer's resources have to be allocated. A computer task is a unit of work that is essential to be performed in order to execute a job. It includes input and output processes, application execution, file management, memory and storage management, device manipulation etc. Task scheduling is crucial for ensuring the right operation of a computer system. An effective task scheduling process gives to the user the impression that the tasks' operation and the computer's response to her/his requests are instant.

Task scheduling is a complex process, that mainly is based on defining the priority of tasks that wait in a queue for execution. The decision about the priority assignment depends on several factors and criteria. Such criteria can be (i) minimizing the mean time that a task waits until its execution, (ii) minimizing the execution time, (iii) minimizing the response time, (iv) minimizing the amount of time it takes to complete a task, (v) maximizing the amount of work that can be completed in a specific period of time, (vi) maximizing fairness in cpu time and device allocation to each task, etc. In addition, the complexity of task scheduling increases in real-time environments, where task scheduling process has to ensure that the tasks' execution is performed within a specified deadline.

Considering the above, we come up with the conclusion that the task scheduling process is complicated and deals with vagueness and uncertainty. Fuzzy logic is a computational technique that can manage effectively the vagueness in task scheduling, defining the tasks' priority in a more realistic way. Below, various scientific works that regard the incorporation of fuzzy logic in task scheduling are presented (Table A.10 in Appendix A).

- In [309], the author developed a fuzzy logic-based algorithm for scheduling non-periodic jobs on soft real-time single processor system. The aim of the proposed algorithm is to identify the task that has to be executed firstly and, also, to define the priorities of the tasks that wait for execution in the queue. The latter process is performed each time a new job is added in the queue. The use of fuzzy logic in task scheduling aims to minimize average delay, number of overdue jobs, and throughput times of jobs. The proposed task scheduling algorithm is based on Mamdani Fuzzy Inference Model. Particularly, two Mamdani FIS are used. The first FIS takes as input the task exterior priority (i.e. the priority that was given in the task from the user), the task waiting time and the task processing time, and gives as output the task processing priority. The output of the first FIS is entered along with the task deadline time as inputs in the second FIS, which gives as output the task final priority. All input and output variables are described through bell-shaped membership functions. Twenty seven fuzzy rules and nine fuzzy rules are used in the first and second madani FIS, correspondingly. The centroid method is used as defuzzification method.
- In [310], the authors present a fuzzy logic-based technique for task scheduling in battery-powered real time systems. The aim of the proposed fuzzy logic-based technique is to reduce energy consumption. by making appropriate decisions on the processor speed/voltage during the scheduling of real time tasks. The following factors are considered for each task that waits to execute: (i) the worst case execution time, (ii) deadline, (iii) blocking time, (iv) arriving time. These parameters are the input values of a Mamdani-type Fuzzy Inference Model, which determines the task priority and the suitable processor speed to execute it. All membership functions of input and output linguistic variables of the proposed system are triangular and trapezoidal.
- In [311], the authors present a fuzzy logic-based approach to make strategic decision about task offloading in Edge Computing Environment. The proposed fuzzy logic-based Inference System takes as input the following five linguistic variables: (i) task size, (ii) network bandwidth, (iii) network delay, (iv) energy consumption, and (v) Mobile Edge Computing Virtual Machine utilization. All

input variables are described through trapezoidal membership functions. 243 fuzzy rules are applied over the input variables and decide if the task will offload in the Mobile Edge Computing server or will be executed locally on the user's device. The defuzzification of the output is performed by using the weighted average method.

- In [312], the authors propose a Type-2 Fuzzy Logic system for managing the uncertainty of task scheduling in an edge computing environment. Particularly, the proposed fuzzy logic-based system contributes to the selection of the most suitable peers for offloading task, eliminating the latency in the delivery of responses and saving the bandwidth of the network. The presented Type-2 Fuzzy Logic system includes three inputs and one output. The three inputs are: (i) the room for execution of additional tasks in a node, (ii) the speed that the task is delivered at pre-defined intervals, (iii) the load that the task will cause to the processing node. The output is the Potential of Allocation. The generation of the presented system's fuzzy sets and membership functions are performed automatically through machine learning and statistical modelling.
- In [313], the authors present a fuzzy logic-based task scheduling technique for real time systems. The aim of the proposed technique is the efficient usage of CPU and minimizing tasks' throughput, waiting time and turnaround time. The proposed system is a Mamdani-type Fuzzy Inference System, which takes as input the task's burst time, response ration and priority, and gives as output the new calculated priority of the task.
- In [314], the authors present a fuzzy logic-based technique for scheduling of workflows on cloud computing environment. The proposed Fuzzy Inference System considers as input for each task (i) the expected execution time, (ii) the assigned priority, and (iii) the task deadline. Then it applies fuzzy rules over the input variables and estimates the overall task priority, which constitutes the criterion for tasks mapping to different type of virtual machines in the cloud environment. The proposed system succeeds better to meet tasks' deadlines.
- In [315], the authors present a fuzzy logic-based task scheduling approach for multiprocessor systems. The aim is to succeed high performance of these systems. The sequence of the tasks is described through a Directed Acyclic Graph, which is used for the calculation of the tasks' priority. Particularly, the proposed technique includes a Mamdani-type Fuzzy Inference System, which takes as input the height of the node, which represents the task, in the Directed Acyclic Graph and the task's execution time. Then, nine fuzzy rules are applied, and the task's priority is calculated as the fuzzy system's output. The input variables of the described fuzzy inference system, as well as its output variable are described trough three fuzzy sets ('Low', 'Medium', 'High'), whose membership functions are triangular.
- In [316], the authors designed a scheduler for multiprocessor real-time systems. The presented scheduler embeds two Mamdani-type Fuzzy Inference Systems. The first FIS, which is called Global Fuzzy scheduler and it is a long-term scheduler, takes as input the job exterior priority, the job processing priority, and the job waiting time and gives as output the job processing priority. The second FIS, which is called Local Adaptive scheduler and it is a short-term scheduler, takes

as input the job processing priority, which is the output of the first FIS and the job worst case execution time, and gives as output the job final priority. The proposed fuzzy logic-based scheduler minimizes the processor time and breakthroughs the limit of the number of total tasks.

- In [317], the authors present a fuzzy logic-based system for real-time tasks scheduling. Particularly, the presented system is a Mamdani-type Fuzzy Inference System, which takes as input the task's arrival time, computation time, and deadline. The input variables are described through trapezoidal membership functions. In more details, the fuzzy sets for arrival time are 'late', 'intermediate', 'late'. The fuzzy sets for computation time are 'short', 'medium', 'long'. The fuzzy sets for deadline are 'critical', 'normal', 'sufficient'. Twenty-seven fuzzy rules of the format 'If arrival time is X and computation time is Y and deadline is Z then priority is P', are applied by the system to generate the task's runtime priority for execution, which is the output of the proposed fuzzy inference system. The system's output is described through three ('low', 'normal', 'high') triangular membership functions. The proposed fuzzy logic-based approach decreases the minimum response time, the turnaround time and the number of deadlines missed.
- In [318], the author describes a fuzzy logic-based task scheduling method, which is based on a Mamdani-type Fuzzy Inference System. The proposed system takes as input the task's deadline and remaining execution time and gives as output the priority for execution. The task's deadline is described through three triangular fuzzy sets, while the task's remaining execution time is described through four trapezoidal fuzzy sets. The system embeds nine fuzzy rules. The presented fuzzy logic-based scheduling technique minimizes the mean waiting time and the mean turnaround time.
- In [319], the author presents a rule-based reasoner that decides dynamically about the priority of the tasks that wait to be executed. The operation of the presented reasoner is based on a Mamdani-type Fuzzy Inference System, which takes as input the task's execution time and waiting time and gives as output the task's priority. Both input variables are expressed through four fuzzy sets ('low', 'medium', 'high' and 'very high'), which are described by trapezoidal membership functions. Also, the system's output variable is described through the following four fuzzy sets: 'Very High', 'High', 'Medium' and 'Low'. The fuzzy rule base of the system consists of eight fuzzy rules.
- In [320], the authors present a task scheduling approach, which combines Q-learning [321] with Ant Colony Optimization [322], for distributes systems. In addition, the presented approach uses fuzzy logic to calculate worker reward value. Particularly, a Mamdani-type Fuzzy Inference System, which takes as input two linguistic variables, the node rapidity and the node state, and gives as output the node reward, is used. For the representation of the node rapidity three trapezoidal fuzzy sets ('speed', 'normal', 'slow') are used. The node state is described through five fuzzy sets ('very overloaded', 'over loaded', 'normal', 'under loaded', 'very underloaded'), which are calculated by triangular and trapezoidal fuzzy sets. The output of the system is described through the following four triangular fuzzy

sets: 'very bad', 'bad', 'good', 'very good'. The system uses seven fuzzy rules to estimate the reward value.

- In [323], the authors present a task scheduling approach for cloud computing environment. The proposed approach is based on meta-heuristic algorithms, which embed fuzzy logic. Particularly, the presented task scheduling approach uses two hybrid algorithms: (i) Particle Swarm Optimization algorithm with fuzzy logic and (ii) Particle Swarm Optimization algorithm with simulated annealing. Fuzzy Logic is applied to adjust the inertia weight and to calculate the fitness value. In more details, two Mamdani-type Fuzzy Inference Systems are used. The first system takes as input the iteration, that is performed until optimize the distribution of resources, and gives as output the inertia weight. Both, system's input and output linguistic variables are described through triangular fuzzy sets. Three fuzzy rules are used for the calculation of the inertia weight. The second system takes as input the task length, the CPU speed, the RAM memory and the status of the virtual machine and gives as output the fitness value. All system's input and output linguistic variables are described through triangular fuzzy sets. Twenty four fuzzy rules are used for the calculation of the fitness value. The proposed approach results in advances in waiting time, queue length, makespan, cost, resource utilization, degree of imbalance, and load balancing.

Regarding the above, the use of fuzzy logic in task scheduling concerns mainly real-time computing systems, but, also, there is an increased scientific interest in application of fuzzy logic in cloud and edge computing environment. Furthermore, the most common used fuzzy technique in task scheduling approaches is the Mamdani-type Fuzzy Inference System. Also, in most fuzzy logic-based approaches the calculated fuzzy output variable is the task priority.

3.6.2 *Fuzzy Logic in Network Traffic Management*

Computer networks are a huge achievement of computer and communication technology research, which provides a breakthrough in communication, allowing exchanging and sharing of data and resources into an organization or worldwide through the Internet. The distribution of data through a network is performed through data packets. In more details, data is broken down in smaller batches, which are called data packets, to manage them more efficiently and send them with more reliability. The distribution of data packets over a network is performed following a variety of communication protocols. However, network congestion can occur either due to high amount of data, which block the network and cause low speeds and/or data loss, or due to malicious attacks. Fuzzy Logic is an effective and intelligent way to monitor and manage a network traffic. Linguistic terms are used to describe congestion parameters and fuzzy rules are used to determine the appropriate action. In the literature review, there are a variety of fuzzy logic application in network traffic management (Table A.11 in Appendix A). For example:

- In [324], the authors used a fuzzy neural network to forecast network traffic. Furthermore, in [325], the authors used ANFIS for botnet detection, which uses the extracted features from the network traffic data. Also, in [326], the authors, used ANFIS to avoid congestion in mobile communication networks. Particularly, the proposed system takes into account three network parameters (received signal strength, received signal Quality, network traffic) and estimates the hysteresis value, which determined the handover process.
- In [327], the authors used fuzzy logic in combination with a genetic algorithm to detect network anomalies. Particularly, the authors developed a fuzzy logic-based expert system, which monitor's the network's traffic and generates alarms when a network problem is detected. The proposed system considers parameters of flows, like bits per second, packets per second, source IP entropy, destination IP entropy, source port entropy, and destination port entropy and calculates thresholds and defines the network traffic characterization. Thresholds and network traffic characterization are used to calculate the anomaly score of the network behavior through a Gaussian membership function. The result of the membership function was compared with a defined threshold and decisions about generating an alarm are taken.
- In [328], the authors presented a fuzzy controller for the calculation of bandwidth consumption. The proposed fuzzy controller is a Mamdani-type Fuzzy Inference system, which takes as input the following three parameters: (i) the average of mean time between two consecutive packets arriving at the interface (ART), (ii) the number of packets of a specific protocol detected (NPP), and (iii) the ratio of number of protocol packets to the total number of packets arriving (PPR). All three input variables are described through trapezoidal membership functions and are taken three linguistic values: 'low', 'medium', 'high'. The same linguistic terms are used for the description of the system's output, which is the percentage bandwidth consumption by each protocol. Twenty-seven fuzzy rules are used for the calculation of the output. These rules are of the following format: " if ART is X and NPP is Y and PPR is Z then bandwidth consumption is W", where X, Y, Z, W ∈ {'low', 'medium', 'high'}.
- In [329], the authors presented a Mamdani Inference System to detect anomaly network traffic in wireless sensor networks. The proposed system has two input variables and one output variable. The input variables are mean and variance of sensory data, which are described by three fuzzy sets ('Low', 'Normal', 'High'). The output fuzzy variable is the estimated type of network traffic, which indicates if the sensory data is normal, suspect or anomaly. It takes five linguistic values, which are described through triangular and trapezoidal membership functions. The rule base of the proposed system includes 9 fuzzy rules of the format 'IF mean is A and variance is B THEN network traffic is C'.
- In [330], the authors proposed a Mamdani fuzzy inference system, which in combination with regression technique performs adaptive sampling. The proposed sampling method helps managing network traffic. More specifically, the presented fuzzy inference system takes as input (i) the current duration value of inter-sampling interval and (ii) the Euclidean distance, which quantifies the amount

of variation in traffic, and gives as output the updated duration of inter-sample interval. The current duration value of inter-sampling interval is described through three fuzzy sets ('Small', 'Medium', 'Large'), the Euclidean distance is described through other three fuzzy sets ('Low', 'Medium', 'High'), and the updated duration of inter-sample interval is described through four fuzzy sets ('Low Increase', 'No change', 'Low Decrease', 'High decrease'). All fuzzy sets of the system have bell-shaped membership functions.

- In [331], the authors used fuzzy logic for controlling congestion in 6LoWPAN networks. Particularly, they proposed a Mamdani-type fuzzy inference system, which takes as input variables the buffer occupancy, the queue length, the packet priority, and the packet size, and gives as output the transmission rate. Both, input and output variables are described through fuzzy sets, which are defined by triangular membership functions. In more details, the input variables 'buffer occupancy' and 'packet size' are described through three fuzzy sets: 'low', 'medium', and 'high', the input variables 'queue length' and 'packet priority' are described through two fuzzy sets: 'low' and 'high', and the output variable is described through five fuzzy sets: 'Reduce Very Low', 'Reduce Low', 'Reduce Medium, 'Reduce High', 'Reduce Very High'. The system uses 12 fuzzy rules of the following format "If 'buffer occupancy' is A and 'queue length' is B and 'packet size is C and 'packet priority' is D, 'transmission rate' is Z".
- In [332], the authors developed a fuzzy logic-based intelligent network traffic management framework, which calculates the router queues size dynamically and directly avoiding high computational consumption and memory resources consumption. The proposed system is a Mamdani-type Fuzzy Inference System, which takes as input the queue error deviation and the Integration of queue error and gives as output the aggregate controlled incoming traffic rate to a router. Both fuzzy input variables are described through the following nine triangular and trapezoidal fuzzy sets: 'Negative Very Large', 'Negative Large', 'Negative Medium', 'Negative Small', 'Zero', 'Positive Small', 'Positive Medium', 'Positive Large', 'Positive Very Large'. Similarly, the system's fuzzy output variable is described through the following nine triangular fuzzy sets: 'Zero', 'Extremely Small', 'Very Small', 'Small', 'Medium', 'Big', 'Very Big', 'Extremely Big', and 'Maximum'. The system's rule base is consisted by 81 fuzzy rules. Also, the proposed system uses the Center of Gravity method to obtain the crisp value of the output variable.
- In [333], the authors proposed a fuzzy logic-based solution to network traffic problem in migrating parallel web crawler. Particularly, they used a Mamdani-type Fuzzy Inference System to calculate the network traffic. The fuzzy input variables of the proposed system are: (i) communication, (ii) bandwidth, and (iii) noise. The system's fuzzy output variable is the network traffic. Input and output variables of the system are described by using three fuzzy sets ('low', 'medium', 'high'), whose membership functions are triangular. The system uses 27 fuzzy rules of the format: "if 'communication' is X and 'bandwidth' is Y and 'noise' is Z, then 'network traffic' is W".

- In [334], the authors use fuzzy logic in combination with weighted sum model, which is a multi-attribute decision making approach, to select the optimal routing path in 6LoWPAN networks. In more details, the proposed fuzzy weighted sum model is applied considering the buffer occupancy, expected transmission count and routing metric of the network and decides about which node is the best parent to be selected in order to avoid congestion.
- In [335], the authors designed a fuzzy logic-based framework for Quality of Service guaranteed in a Mobile Ad hoc Network, which is self-configuring and connects independently located mobile nodes via wireless links. The proposed framework is based on a Mamdani-type Fuzzy Inference System, which has four input variables (source packet delivery ration, packet QoS type, queue length and packet expiry time). The system calculates the packet's priority index. The fuzzy input variables are described through three fuzzy sets with linguistic values: 'low', 'medium' and 'high'. The system's output variable is described through five fuzzy sets with linguistic values: 'very low', 'low', 'medium', 'high', 'very high'. Fuzzy sets of both, input and output variables, are defined by Gaussian membership functions.

Regarding the above, the use of fuzzy logic in network traffic concerns mainly managing bandwidth consumption and avoiding congestion. Furthermore, the most common used fuzzy technique in network traffic is the Mamdani-type Fuzzy Inference System, while the input and output variables vary.

3.6.3 Fuzzy Logic in Computer Security

Computer security includes the protection of standalone computer systems, computer networks, computer resources and data from malicious software, cyber-attacks, suspicious and no-trust activities, unauthorized use etc. The aim of computer security is to ensure the confidentially, integrity and availability of data. A major category of computer security is network security. Network security is critical for protecting a computer network, its infrastructure, resources, and the data that pass through it from unauthorized access, cyber-attacks, malware, misuse, or other malicious activities.

Computer security includes practices and policies that ensure the secure and trustworthy operation of a standalone computer system or network. These practices, combines, usually, hardware and software solutions. The most known practices of security are: access control, malware protection through antivirus and anti-malware software, firewalls, intrusion detection and prevention systems, encryption, virtual networks, network segmentation, monitoring, backup activities etc.

Given that fuzzy logic is a method for managing uncertainty and ambiguity, it is selected, in many cases of computer security. Fuzzy logic provides flexibility, adaptivity and comprehensiveness. Fuzzy logic can be applied for computer security issues as follow:

- To compute and estimate the risk of no proper activities.

- To express rules of an intrusion detection system in a more flexible and human-like reasoning way, dealing with the uncertain and vague characteristics of suspicious activities or attacks.
- To describe better access permissions and user behaviors.
- To provide a more comprehensive way and allow less strict processes of anomaly detection practices.
- To consider multiple factors, which can be characterized by subjectivity and uncertainty or imprecision, in security decision making processes.
- To provide adaptivity in network security mechanisms modeling dynamically network parameters.

In most circumstances, fuzzy logic is not used as a standalone approach for computer security, but it is embedded in other security mechanisms or approaches. In the literature review, there are a variety of fuzzy logic application in computer security (Tables A.12 and A.13 in Appendix A). Below, recent research works that consider the use of fuzzy logic in computer security issues and practices are presented. They are divided into two categories: network security and computer security other than networks.

Case studies of fuzzy logic use in network security

- In [336], the authors propose an intelligent framework for threat detection in blockchain-based IoT environments. The presented framework includes a novel fuzzy deep learning model, an adaptive neuro-fuzzy inference system (ANFIS), fuzzy matching, and fuzzy control system. Also, metaheuristic algorithms are used for optimization.
- In [337], the authors present an anomaly detection technique on the Internet of Medical Things infrastructure and networks. The proposed approach combines Fuzzy C-Means clustering along with customized Bi-LSTM technique. Fuzzy C-Means clustering is used for processing multi-modal sensory medical signals and extract features. Then, Bi-LSTM technique is used for identifying attack patterns.
- In [338], the authors, use an Adaptive Neuro-Fuzzy Inference System (ANFIS) to estimate the security index in a Vehicular Ad hoc NETwork (VANET), which is responsible for the communication between vehicles. The proposed security model is a prediction model.
- In [340], the authors propose an approach for improving security and stability in vehicular ad hoc network. The proposed approach is based on a fuzzy neural network, which calculates the node stability and trust values of each node of the network. The presented fuzzy neutral network is part of a routing protocol.
- In [341], the authors present a trust model, which incorporates fuzzy logic, to secure the vehicular network. The proposed model includes five steps: (i) sender authentication checking, (ii) sender's experience level measurement, (iii) sender's plausibility level measurement, (iv) message's accuracy level measurement, (v) message's trust value estimation. Fuzzy logic is used for the calculation of the experience level, which is described through three trapezoidal fuzzy sets ('low', 'medium', 'high'), the plausibility level, which, also, is described through three

trapezoidal fuzzy sets ('low', 'medium', 'high'), and the accuracy level, which is described by three fuzzy sets ('low'-trapezoidal, 'medium'-triangular, 'high'-trapezoidal). Furthermore, the proposed approach uses a Mamdani Fuzzy Inference System, that takes as input the experience level, the plausibility level, and the accuracy level, and gives as output the trust value. The incorporated Mamdani Fuzzy Inference System uses 36 fuzzy rules (27 in cases where fog node is available and 9 in cases where fog node is not available). According to the trust value, the system decides about the acceptability or not of the message.

- In [343], the authors use an Adaptive Neuro-Fuzzy Inference System to detect attacks packets. The proposed ANFIS takes as input eight features from the network data, which have been selected by applying feature selection algorithms, use Sugeno fuzzy model for the layer of reasoning and gives as output a crisp number that declared if network data is normal or malicious.
- In [344], the authors designed a security protocol, which tackles back hole attacks in opportunistic networks. The proposed security protocol combines fuzzy logic with game theory. Fuzzy logic is used to characterize a node as malicious or benign. In more details, a Mamdani-type FIS is used. The FIS take as inputs node's parameters (delivery time, fast division, and potential threat count), uses six fuzzy rules, and estimates how malicious the node is. In this way, the definition of nodes' character is based on dynamic, not fixed parameters.
- In [345], the authors present a fuzzy logic-based approach for detecting malicious activities in internet of things systems. Particularly, they combine Bayesian possibilistic clustering module for feature processing with a fuzzy classifier backend, which is consisted of a collection of fuzzy decision trees that provide readable decision boundaries.
- In [347], the authors built a neuro-fuzzy network to detect phishing websites in fog networks. The presented neuro-fuzzy network has five layers, 6 inputs and one output. The 6 inputs regard features of a URL and are: primary domain, subdomain, path domain, page rank, alexa reputation, google index. The output is a number, whose value (if it greater than a threshold or not) determines of the URL is phishing or legitimate.
- In [348], the authors proposed a fuzzy logic-based detection system for jamming attacks in wireless sensor networks. The proposed approach combines a Takagi–Sugeno FIS with ANFIS. The FIS is used to optimize the jamming detection metrics and ANFIS is used to detect various types of jamming. Particularly, FIS takes as input the packet delivery ratio and the received signal strength indicator and calculates the jamming cut-off. Then, ANFIS is used to detect the presence or not of jamming.
- In [349], the authors proposed an insider attack detection approach for wireless multimedia sensor networks. The proposed approach is based on a Mamdani-type FIS, that take as input six parameters: network parameters ('transmission rate', 'forwarding delay', 'packet delivery ratio') and physical parameters ('residual energy', 'transceiver range', 'received signal strength'). The output of the system is the degree of suspect for each node of the network.

- In [350], the authors proposed a fuzzy logic algorithm that calculates the trust level of nodes of a mobile adhoc network (MANET). The node trust level is estimated taking into account energy consumption, the total trust value, the packet veracity check value, and the final trust value. The proposed fuzzy logic-based approach succeeds to detect black hole attack in MANET and provide more reliable security in military data communication.
- In [351], the authors presented a fuzzy logic-based approach for malicious node detection in wireless multimedia sensor network. The proposed approach is based on a Mamdani-type FIS, which takes as input the message success rate and the elapsed time of a sensor and applying nine fuzzy rules over them, estimates the trust value of the node. Both fuzzy input variables can take as value three linguistic values ('low', 'medium', 'high'), which corresponds to fuzzy sets and are described trough trapezoidal and triangular membership functions. The output of the system (trust score) is described through five trapezoidal and triangular fuzzy sets ('lowest', 'low', 'medium', 'high', 'highest'). After detecting the malicious nodes, a security-aware multipath routing, which includes Improved Elliptic Curve Cryptography algorithm, is performed.
- In [352], the authors suggested a fuzzy logic-based approach for detecting sybil attacks in vehicular ad hoc networks. Particularly, they used two Mamdani-type FIS in combination with arithmetic optimization algorithm. Both FIS calculate the security status. However, the first FIS takes as input: (i) vehicles trust, (ii) received signal strength indication difference, (iii) vehicle distance, and (iv) vehicle angle. The second FIS take as input: (i) signal-to-noise ratio, (ii) network entry time, (iii) number of the neighbors, and (iv) buffer size. The outputs of the two FIS are aggregated. The arithmetic optimization algorithm is used to tune the fuzzy sets and to select the rules of the two FIS.
- In [355], the authors developed a multi-level random forest algorithm using a fuzzy inference system for intrusion detection for Internet of Things Networks. Initially, feature selection through a genetic search algorithm is performed. The selected features constitute the input of the proposed random forest algorithm for detecting intrusions. Then, a Mamdani-type FIS is used for the classification of the intrusions as normal, low, medium or high. The FIS takes as input four fuzzy variables ('probing', 'denial of service', 'user-to-root', and 'remote-to-local'), whose values declares the degree of the intrusion of each category. The FIS applies 16 fuzzy rules.
- In [356], the authors use fuzzy Analytical Hierarchy Process (FAHP) to improve the information security control analysis for cloud computing networks and wireless sensor networks. The aim of FAHP is to select the most relevant set of information security controls to satisfy the information security requirements of an organization. The criteria of the proposed FAHP are: (i) implementation time, (ii) effectiveness, (iii) risk, (iv) budgetary constraints, (v) exploitation time, (vi) maintenance cost, and (vii) mitigation time. The weights of the criteria and the alternatives are evaluated by using triangular fuzzy sets.

Case studies of fuzzy logic use in computer security other than networks

- In [339], the authors developed an interval type-2 fuzzy logic controller for risk assessment in cyber security. The proposed controller includes three Mamdani-type Fuzzy Inference systems. The first system takes as input three parameters ('intent', 'targeting', 'capabilities') and gives as output the 'overall capabilities', which is inserted as input into the second fuzzy inference system with the parameter 'vulnerabilities. The output of the 2nd sub-system is the 'overall likelihood', which is inserted as input in the third fuzzy inference system along with the parameter 'impact'. The 3rd and last fuzzy inference sub-system gives as output the risk for cyber-attack or malicious activity.
- In [342], the authors combined fuzzy logic with a variety of machine learning classifiers to detect malware to the Android operating system. Particularly, they used six different machine learning techniques: (i) logistic regression, (ii) Bayes point machine, (iii) boosted decision tree, (iv) neural network, (v) decision forest, and (vi) support vector machine. The outputs of these six classifiers are combined through voting and routing processes. This has as a result three weighted outputs of the machine learning classifiers to be inserted as input in a Mamdani-type Fuzzy Inference System, which estimates the degree in which data is malicious. The three fuzzy input variables of the Mamdani FIS can take five linguistic values ('very small', 'small', 'medium', 'big', 'very big'), which are described through generalized bell-shaped membership functions. The output of the proposed fuzzy inference system is the degree of malware and is described through 13 fuzzy sets, which are defined by generalized bell-shaped membership functions. The rule base of the system is consisted of 125 fuzzy rules.
- In [346], the authors proposed an integrated fuzzy logic-based three-tier trust management framework for mobile edge computing security. The core of the proposed framework is a Mamdani-type FIS, which takes as input three parameters ('Service Level Agreement', 'processing performance', 'violations measurements') and estimates the overall trust of the system. The values of the input variables are calculated in a previous phase. They are described by three fuzzy sets, whose linguistic values are 'low', 'medium', 'high', and are defined by trapezoidal membership functions. The proposed FIS applies 21 fuzzy rules over the input variables and gives as output the overall trust, which is described using four triangular fuzzy sets, namely 'low', 'medium', 'high', 'excellent'.
- In [353], the authors developed a fuzzy rule-based expert system that evaluates trustworthiness of a cloud service provider. The presented system included four Mamdani-type FIS. The outputs of the three FIS constitute the input variables for the fourth FIS, which estimated the final trust value of the provider. In more details, the first FIS takes as input four fuzzy variables ('reliability', 'performance', 'security', 'usability'), whose values are defined by the ratings of consumers. This system's output is called 'direct-trust' and depicts the provider's trustworthiness from the viewpoint of consumers. The second FIS takes as input the same variables with the first FIS, but their values are defined by the average ratings of the four parameters given by all the consumers to different cloud service providers.

The output of the second FIS is 'public-review' and concerns all the cloud server providers. The third FIS takes as input four fuzzy variables ('performance', 'availability', 'scalability', 'accuracy'), whose values are defined after the results of the monitoring by the auditor. The third system's output is called 'auditor-trust' and depicts the provider's trustworthiness from the viewpoint of auditor.

- In [354], the authors proposed a fuzzy logic-based approach for intrusion detection in computer security regarding the user behavior. The approach is based on the building of a finite automaton, which models the user's normal behavior. Fuzzy logic is used, then, for threat evaluation. Particularly, three Mamdani-type FIS are used as a reasoning system for describing human knowledge. The combination of these FIS estimates how normal or not is the user's behavior. In more details, the first FIS takes as input the number of match and the number of match continuity, and estimates the degree of match of user's behavior to the expected and normal behavior. Similarly, the second FIS takes as input the number of mismatch and the number of mismatch continuity, and estimates the degree of mismatch of user's behavior to the expected and normal behavior. Then, the output of the two aforementioned FIS are inserted into the third FIS, which calculated the degree of threat in the user's behavior.
- In [357], the authors proposed a security framework that embeds semi-supervised and fuzzy based learning model. The aim of the proposed fuzzy logic-based system is to improve the efficiency of an intrusion detection system classifier for protecting data. To succeed it, the authors combine fuzzy logic with machine learning.
- In [358], the authors suggest a fuzzy logic aided intelligent threat detection solution for cisco adaptive security appliance 5500 series firewalls. The presented fuzzy logic-based system informs about the detected threats the administrators and users in a comprehensive way. It is based on a Mamdani-type FIS, which takes as input statistics values about denial by access control list, SYN Attack Detected and DoS attack detected, and gives as output the overall security threat. All input and output variables of the system are described through three triangular fuzzy sets ('low', 'medium', 'high').
- In [359], the authors present an anomaly detection system that operates on cloud and ensure the privacy and quality of the data in the infrastructure of smart cities. The proposed system applies homomorphic encryption and MapReduce in combination with Fuzzy C-Means.
- In [360], the authors combine fuzzy logic with machine learning techniques to increase the detection rate and decrease the false positives rates of intrusion detection systems. In more details, fuzzy-c means is used, firstly, to cluster the instances of data samples as normal or anomalous. Then, the instances, which was characterized as anomalous, are inserted into the next phase, where further detection of anomaly activities is performed through K nearest neighbor algorithm.
- In [361], the authors used a fuzzy logic-based algorithm to classify anomaly activities in intrusion detection systems. Particularly, they use Fuzzy Kernel Robust Clustering algorithm to classify the instances of a data set into five clusters that model the activities. These clusters are: normal, denial of service, probe, user

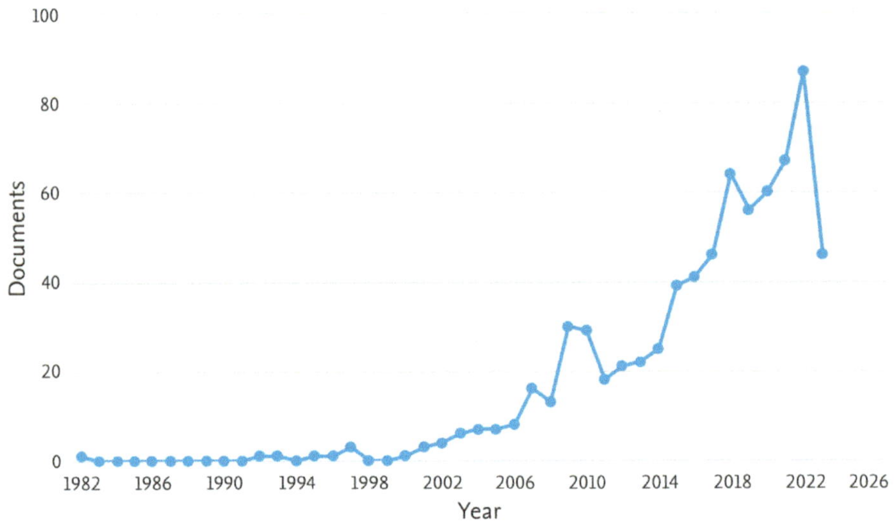

Fig. 3.6 Statistics concerning the published studies in fuzzy logic in computer and network security

to root, and remote to local. The classification is performed considering 41 data features.

- In [362], the authors use an evolving fuzzy rule-based approach for real-time anomaly detection in data centers. Particularly, the proposed approach is a Gaussian Fuzzy Classifier, which is built on the fly concerning attributes and summaries of log records.

Considering the above, we notice that fuzzy logic is used, especially, in network security. Fuzzy techniques that are preferred to be used for protection of intrusion and attacks are: Fuzzy neural networks, Fuzzy Inference Systems, ANFIS, and Fuzzy C-Means. According to a search (it is conducted July 18, 2023), which was carried out in 'Scopus' scientific database, with keywords "fuzzy logic" AND ("computer security" OR "network security") in article's title, abstract or keywords, the first research work regarding fuzzy logic and computer security was published in 1982. From then, there is an increased scientific interest in this field (Fig. 3.6). In Fig. 3.7, the number of the published research documents of the last ten years (from 2013 to 2023) in the field of fuzzy logic-based computer security systems are presented per fuzzy technique that they describe.

3.7 Fuzzy Logic-Based Navigation Software Systems

A navigation system is a system, which helps drivers and autonomous vehicles in navigation. Particularly, a navigation system can:

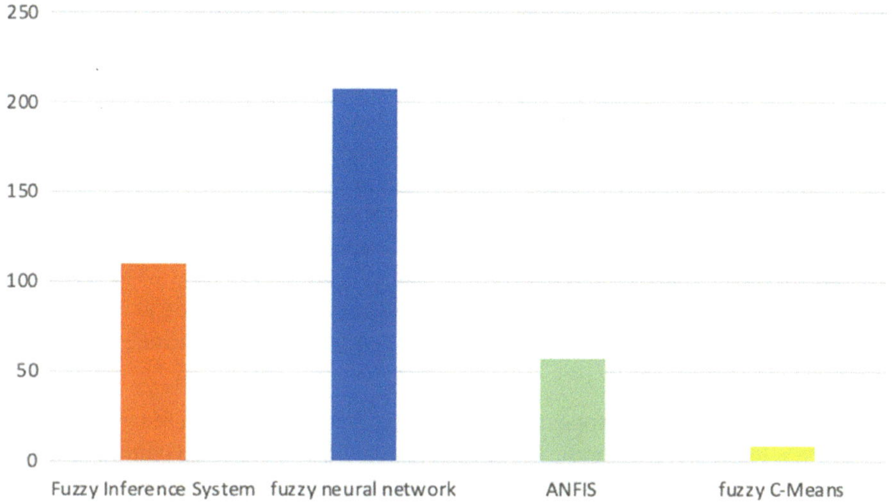

Fig. 3.7 2013-2023: number of published research documents concerning computer and network security per fuzzy logic-based technique

- detect and determine a vehicle's location, velocity and/or orientation,
- provide suggestions concerning the vehicle's motion,
- provide directions,
- provide information about obstacles, or dangerous driving conditions.

A navigation system can be used in cars, tracks, ships, underwater vehicles, planes, autonomous vehicles, robots etc. There are several types of navigation systems. The main types of them are:

- Global Navigation Satellite System (GNSS): is a general term describing any satellite constellation that provides positioning, navigation, and timing services on a global or regional basis [363].
- Global Positioning System (GPS): is a navigation system using satellites, a receiver and algorithms to synchronize location, velocity and time data for air, sea and land travel [364].
- Inertial Navigation System (INS): is a navigation device that uses motion sensors, rotation sensors and a computer to continuously calculate by dead reckoning the position, the orientation, and the velocity of a moving object without the need for external references [365].
- Strapdown Inertial Navigation System (SINS): is comprised of accelerometers and gyroscopes that measure the motion of the aircraft and provide information about the direction the aircraft is facing [366].
- Automotive Navigation System.
- Marine (or Ship) Navigation System.

The operation of a navigation system is based on sensors, cameras, digital maps, satellite signals, image processing techniques. The sensors data, signal processing and image processing include, often, noisy and imprecise data. Furthermore, the differences in driving motor and manufacturing characteristics of vehicles, the variant driving and road conditions and the several physical phenomena, like wind, rain, snow and waves increase the uncertainty of a navigation system. An efficient way to deal with this uncertainty is the use of fuzzy logic in a navigation system. By integrating fuzzy logic in a navigation system, the navigation accuracy, reliability and robustness are improved.

In the literature review, there are several studies, which concern the use of fuzzy logic in navigation systems (Table A.14 in Appendix A). Below, studies of the last recent years are presented:

- In [367], the authors proposed a knowledge-based fuzzy inference system to enhance long-term performance of concessional SINS/GPS navigation systems, for which the orientation accuracy is the main purpose. The proposed fuzzy logic-based system is a Mamdani type Fuzzy Inference System, which allows the efficient cooperation between attitude-heading reference system (AHRS) and strap-down inertial navigation system (SINS). The fuzzy logic-based system takes as input the inertial measurements comprising gyroscopes and accelerometers and gives as output the maneuvering level of the vehicle.
- In [368], the authors proposed a hybrid fuzzy method for performance evaluation of fusion algorithms for integrated navigation system. The presented hybrid fuzzy method combines fuzzy logarithmic least square AHP method (fuzzy LLSM) with fuzzy comprehensive evaluation (FCE) method. The proposed method deals efficiently with uncertainty and subjectivity.
- In [369], the authors combine type-2 fuzzy sets with Hammerstein neural network to improve the navigation accuracy of low-cost SINS in combination of global navigation satellite system (GNSS).
- In [370], the authors use a type-2 fuzzy logic system in cooperation with extreme learning machine to improve the position speed and accuracy in high-speed vehicular navigation system. Extreme learning machine is used to generalize efficiently the performance and fast learning speed. The proposed type-2 fuzzy logic system is used to manage the uncertainty that exists in GPS/INS data, which constitute the input data of the extreme learning machine algorithm.
- In [371], a vision navigation system, which is based on a fuzzy control algorithm. The presented fuzzy control algorithm is a Mamdani-type Fuzzy Inference System, which combines the navigation deviation parameters obtained from image processing with the motion and control characteristics of the Automated Guided Vehicle and calculates the virtual wheel deflection angle and the velocity of each active wheel. The system uses seven triangular fuzzy sets to describe the parameters of vehicle's motion. Also, the system uses 49 fuzzy rules to take decision about the vehicle's guidance. In such way, the navigation system allows the Automated Guided Vehicle to move along the road marking lines.

- In [372], a Sugeno Fuzzy Inference system, which is integrated in a docking guidance system of an autonomous underwater vehicle, is presented. The aim of the proposed fuzzy inference system is to navigate and control the autonomous underwater vehicle into the docking station in a strategic manner and execute different maneuvers. The system takes as input the approach range and the cross range and applying fuzzy rules, in the form of functions, over them determines the desired vehicle speed and heading commands for guidance. The fuzzy system's output variables are sent to the low-level vehicle controllers for execution. The system's fuzzy inputs are described through triangular and trapezoidal membership functions.
- In [373], the authors integrate an adaptive neural fuzzy inference system (ANFIS) into a SINS/USBL (ultra-short baseline) navigation system for underwater vehicles. USBL is a method of underwater acoustic positioning, which deal with vague and imprecise information. Therefore, the use of ANFIS enhances the navigation accuracy. The fuzzy rules of the proposed ANFIS are based on Takagi–Sugeno Fuzzy Inference System and for the fuzzification of the input generalized bell-shaped membership functions are used. In this study, the ANFIS is combined with variational Bayesian Kalman filter and principal component analysis.
- In [374], the authors present a fuzzy logic-based approach for verification of indoor navigation systems. The proposed approach is based on a neural-fuzzy network, which determines the system's current position and identifies possible errors.
- In [375], an adaptive fuzzy logic-based filter for enhancing the navigation accuracy and robustness of the visual inertial odometry navigation systems is presented. The authors used a Takagi–Sugeno Fuzzy Inference System, which takes as input the divergence degree of visual inertial odometry filter and gives as output a scalar factor, which contributes to the enhancement of the robustness and the estimation accuracy of the visual inertial odometry navigation system. The fuzzy values of the system's input variable are expressed through the following four fuzzy sets: 'normal', 'a little big', 'big', 'very big', which are described through triangular and trapezoidal membership functions. The fuzzy system uses four 'IF-THEN' format fuzzy rules to estimate the scalar factor as a function of the divergence degree.
- In [376], the authors present a fuzzy logic-based expert system, which is integrated in a loosely-coupled navigation system, whose operation is based on satellite navigation and low-cost inertial MEMS sensors. The fuzzy logic-based expert system is a Mamdani Fuzzy Inference System, which takes as input the following variables: the rotation state, the accelerometer readings, the wheels' rotation speed. The system gives as output two weight coefficients that are used for the calculation of pitch and roll angles. For the representation of the input and output fuzzy variables, triangular membership functions are used. The aim of the presented system is to identify the movement state and type.
- In [377], a type-2 fuzzy inference system that enhances the positioning accuracy, performance, and reliability of low-cost SINS/GNSS-integrated navigation system, is presented. The system takes as input the maneuvering level of the

vehicle and the inertial measurement units' measurements. The input variables of the system are described through three fuzzy sets, whose membership functions are triangular and trapezoidal. The system applies a set of nine fuzzy rules over the input and estimates a normalizing factor, which is used to automatically adjust the system's noise covariance matrix.

- In [378], the authors combine adaptive neuro-fuzzy inference system (ANFIS) with subtractive clustering algorithm to improve the accuracy of the prediction of the vehicle location. The integrated ANFIS uses rules that are based on Takagi–Sugeno Fuzzy Inference System. The subtractive clustering algorithm contributes to use in ANFIS the optimum and fewer number of membership functions and fuzzy rules. The proposed approach improves navigation and reduces processing time and memory for prediction accuracy of the vehicle's position.
- In [379] a Mamdani Fuzzy Inference System is used to define the reward of the vehicle depending on its current state. The reward indicates if a driving action should be reinforced, avoided, or have a neutral response. The system encourages driving in the center of a raceway and with low to moderate speed when the vehicle is in a curve route and with moderate to high speed when the vehicle is in a straight path. The presented fuzzy inference system takes as input the vehicle's speed, steering angle, and distance from the center of the track and gives as output the reward. The input and output linguistic variables of the system are represented through Gaussian membership functions.
- In [380] a Takagi–Sugeno Fuzzy Inference System is used to enhance the navigation estimation accuracy of GPS/INS navigation systems. The aim of the proposed fuzzy inference system is to adapt the noise strength in the internal model of the Kalman filter. Particularly, it takes as input two degree of divergence parameters, whose linguistic values are described through triangular membership functions. The system's output is used to adjust the weighting factor of the process noise covariance matrix of the Kalman filter.
- In [381], the authors used fuzzy inference system to decide about the existence or not of a roundabout. The presented fuzzy system is a Mamdani Fuzzy Inference System, which takes as input the right curb, the right curb, and the elliptical curve. These data is derived from image that is captured from a camera and a laser range finder. Four fuzzy rules are applied over the input variables and give as output the value of a linguistic variable, which is called conditions and defines the existence or not of a roundabout. The input and output linguistic variables of the system are described through trapezoidal membership functions.
- In [382], the authors develop a navigation system for recognition and tracking of autonomous underwater vehicles. The proposed system is based on color recognition, fuzzy logic controllers and Kalman filtering. The fuzzy logic controller is based on Tsukamoto fuzzy model. It takes as input the image area and coordinates and gives as output the rudder angle of the servomotor and the propeller revolution speed.
- In [383], an adaptive neuro fuzzy inference system (ANFIS), which improves ship navigation considering wind, current and waves is presented. The proposed ANFIS takes as input the wind speed, the wind direction, the wave height and the

wave direction, which are fuzzified through Gaussian membership functions, and give as output the ship's speed change.

- In [384], the authors designed a Fuzzy Logic Controller for ship steering control. The proposed Fuzzy Logic Controller is based on Mamdani-type Fuzzy Inference System. It takes as input the yaw error and yaw rate of the ship and gives as output the ship's rudder angle. The input and output linguistic variables of the proposed fuzzy system are described through five and seven fuzzy sets accordingly, which are expressed through triangular and trapezoidal membership functions.
- In [385], the authors propose a fuzzy logic-based navigation algorithm for autonomous vehicles, which is based on RFID sensors. The presented fuzzy system is a Mamdani fuzzy Inference System that takes as input the received intensity indicator signal levels from the RFID for the left and right antennas and gives as output the time duration of a pulse used for pulse position modulation to command a servo motor. The input and output linguistic variables of the system are described through five fuzzy sets, whose membership functions are Gaussian.
- In [386], the authors propose a knowledge-based neural fuzzy controller for mobile robot navigation and dead ends avoidance. The fuzzy logic-based controller takes as input the distance between the robot and the obstacle, the left sensor, the right sensor, and the front sensor, and gives as output the velocity of the two wheels of the mobile robot.
- In [387], the authors present a 5-layer Adaptive Neuro-Fuzzy Inference System (ANFIS) to optimize the navigation of a mobile robot. It accepts as input the data measured by the sensors of the robot and transforms them in linguistic values, which are described by fuzzy sets whose membership functions are Gaussian. Also, the presented ANFIS uses 24 fuzzy rules, which are based on Takagi–Sugeno Fuzzy Inference System. It calculates the robot's steering angle. In the proposed approach ANFIS is combined with Particle Swarm Optimization.
- In [388], a fuzzy logic-based controller, which helps a mobile robot to avoid obstacles is proposed. The operation of the presented fuzzy logic-based controller is based on an adaptive neuro-fuzzy inference system (ANFIS). The input of the proposed ANFIS includes the sonar sensors data. Particularly, the left, right and front distance of the robot from the obstacle are fuzzified into three fuzzy sets: 'near', 'mid', far', through bell-shaped membership functions. Also, 27 Sugeno type if–then fuzzy rules are used. The proposed ANFIS estimates as output the desired steering angle.

Considering the above, we notice that there is an increased scientific interest in automotive navigation systems. Furthermore, Sugeno Fuzzy Inference Systems is preferred to be integrated in navigation systems. This is derived from a search, which is performed in academic web-based database 'Scopus'. Particularly, a search, which was carried out in 'Scopus' with keywords "fuzzy" AND "Mamdani" AND "Navigation" in article's title, abstract or keywords, gave as output 44 documents from 2013 to 2023 (Fig. 3.8). Similarly, a search, which was carried out in 'Scopus' with keywords "fuzzy" AND "Sugeno" AND "Navigation" in article's title, abstract

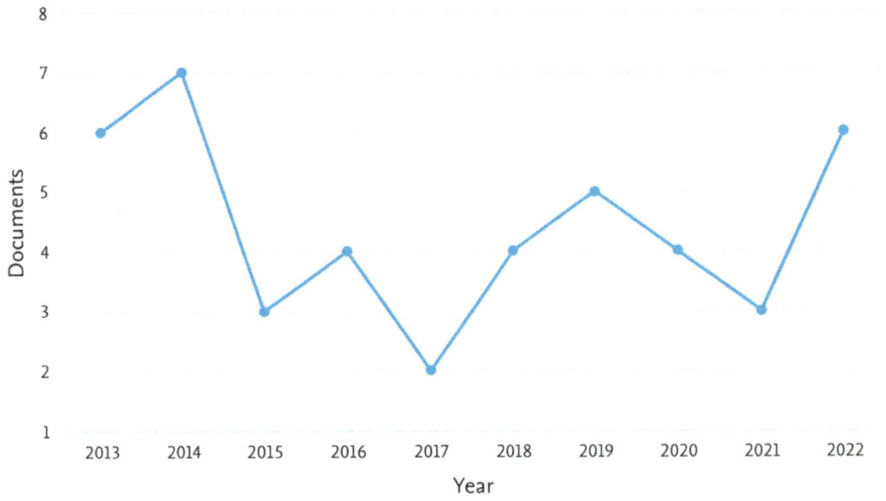

Fig. 3.8 Statistics concerning the published studies in fuzzy logic-based navigational systems, which use Mamdani-type FIS

or keywords, gave as output 192 documents from 2013 to 2023 (Fig. 3.9). Furthermore, ANFIS are used extensively in navigation systems, especially in navigation systems of mobile robots. A search, which was carried out in 'Scopus' with keywords "ANFIS" AND "Navigation" in article's title, abstract or keywords, gave as output 146 documents from 2013 to 2023 (Fig. 3.10). From them, 57 (39%) documents concern mobile robots.

3.8 Fuzzy Logic-Based Heating, Ventilation and Air Conditioning (HVAC) Systems

Nowadays, the improvement of technology allows the provision of thermal comfort and healthy air quality in apartments, houses, budlings, industrials, offices, and general, in enclosed spaces. This is achieved through systems which control the temperature, humidity, and air quality. Such systems are called heating, ventilation, and air conditioning systems (HVAC). They include radiators, heat pumps, electric heaters, fan heaters, boilers, fans, air handling units, air conditions, ventilators, humidifiers, dehumidifiers, thermostats, filtering systems etc. Synchronous HVAC systems embed technologies that enable a more smart and effective controlling of environmental conditions parameters. A such technology is fuzzy logic, which allow the handling of uncertainty and imprecision that characterize the air quality, temperature and humidity controlling. Fuzzy logic enhances HVAC systems' control and decision-making processes, improves their performance, makes them more energy

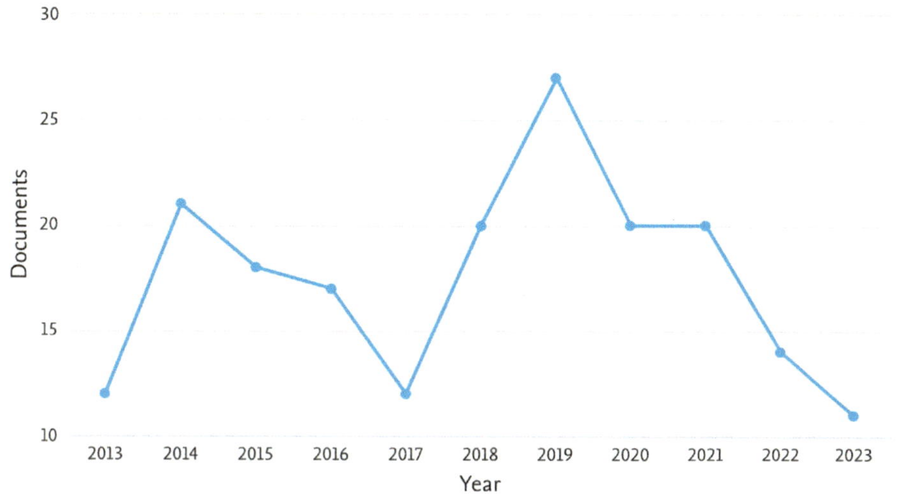

Fig. 3.9 Statistics concerning the published studies in fuzzy logic-based navigational systems, which use Sugeno-type FIS

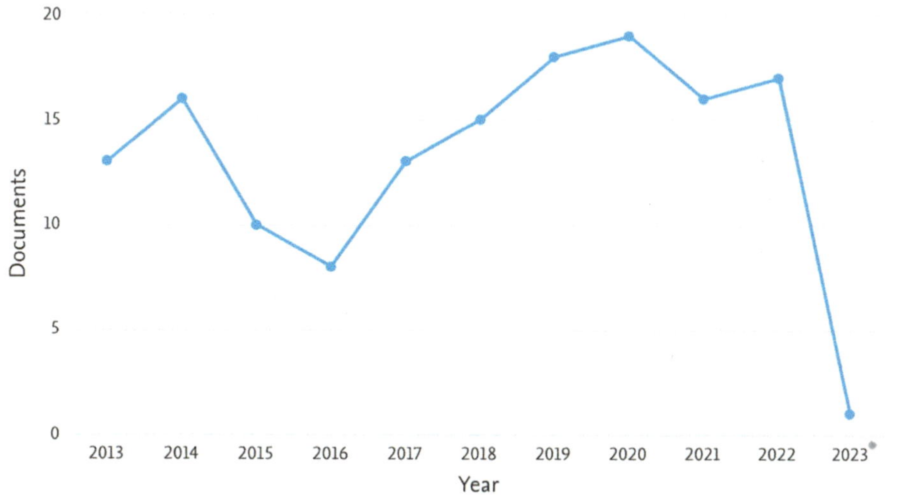

Fig. 3.10 Statistics concerning the published studies in fuzzy logic-based navigational systems, which use ANFIS

efficient and enables them to operate in a more sophisticated and adaptive way. Particularly, fuzzy logic is used to HVAC systems to the following processes:

- Temperature control: the fuzzy logic allows a more adaptive and effective controlling of the temperature considering multiple parameters of the environment conditions.

- Humidity control: the fuzzy logic allows a more adaptive and accurate controlling of the humidity considering multiple parameters of the environment conditions.
- Airflow control: the fuzzy logic allows a more effective determination of the fan speed and/or damper position considering several variables of the environment conditions, succeeding delivery of the right amount of conditioned air.
- Energy efficiency: Fuzzy logic contributes to more efficient management of energy consumption by dynamically adjusting of HVAC system's operations.
- Fault detection and diagnosis: Fuzzy logic enables the monitoring of several HVAC system's operations and parameters. It allows the identification of system's deviations from expected values and behavior and their analysis in more explainable and accurate way. Therefore, early fault detection is achieved, and timely system's maintenance or repair is enabled having as a result future failures prevention and system's performance improvement.
- System's adaptability: Fuzzy logic considers user's preferences, environmental conditions, feedback and comfort models and adjust HVAC system's operations and output parameters accordingly.

In the literature review, there are a variety of fuzzy logic application in HVAC systems. Below, recent research works that consider the use of fuzzy logic in heating, ventilation and air conditioning (Table A.15 in Appendix A) systems are presented.

- In [389], the authors developed a smart fuzzy logic-based system for energy management, which adjusts autonomously the thermostat's temperature considering energy prices, environmental conditions, users' schedule and preferences. Particularly, the presented system embeds an adaptive fuzzy logic model, which detects, learns, and adapts to current user's preferences. The proposed adaptive fuzzy logic model is based in a fuzzy logic rule-based system, which takes into consideration the house parameters, the house's thermal characteristics, outdoor/indoor temperatures, electricity prices, occupancy, changes in user preferences and schedules, and gives as output the temperature in which the system has to be set. System's input data is received from sensors and are described through triangular and trapezoidal membership functions. The decision-making process is based on 27 fuzzy rules.
- In [390], the authors present a technique, which combines an adaptive-network based fuzzy inference system with back-propagation network based artificial neural network, to forecast meteorological data for thermal systems. The proposed ANFIS takes as input variables months, number of sky clear days, cloudiness, average air pressure, ground surface minimum temperature, mean solar time, and evaporation and using Sugeno-type fuzzy rules, calculates the temperature, solar radiation, and relative humidity.
- In [391], the authors developed a personalized air conditioning thermal controller, whose operation is based on a fuzzy logic estimation model. In more details, the fuzzy logic-based model is a Mamdani-type Fuzzy Inference System, which takes as input the indoor air temperature and the occupants' actual thermal sensation and gives as output the set point temperature and the fan speed. 135 fuzzy rules are used for the calculation of the output variables. The input and output variables

take linguistic values, which are described through triangular and trapezoidal membership functions.

- In [392], the authors present a mechanism for temperature regulation in HVAC systems with low computational complexity. The proposed mechanism combines two Mamdani-type Fuzzy Inference Systems with artificial neural network. The artificial neural network is used to compute the temperature set-points per thermal zone. The two FIS are used for energy consumption management. Particularly, the first FIS takes as input the weather data, the available funds and the current market's energy trading rate and estimates the availability of energy. The second FIS takes as input the occupants' thermal comfort and the thermostat's temperature set-point, as it has been calculated by the artificial neural network, and estimates the current demand for energy consumption.
- In [393], the authors developed a smart thermostat, which is combined with a home energy management system. The developed system incorporates a Mamdani-type FIS, which is responsible the demand response of air-conditioning and maintenance of thermal comfort. In more details, the proposed FIS takes as input three parameters: (i) electricity prices, (ii) solar radiation, and (iii) occupant presence, and applying over them 27 fuzzy rules calculated dynamically the set point temperature change. For all input variables, three fuzzy sets ('low', 'medium', 'high') with trapezoidal membership functions are used. For the system's output five fuzzy sets are used: 'negative big' (trapezoidal), 'negative small' (triangular), 'zero' (triangular), 'positive small' (triangular), 'positive big' (trapezoidal)
- In [394], the authors present a smart approach for energy consumption reduction of HVAC systems remaining the occupants' thermal comfort. The proposed system is based on fuzzy logic and machine to machine communication. Machine to machine communication is used to exchange the temperature of all of the office neighbors of a building. Then, a Mamdani-type Fuzzy Inference System is used to calculate the temperature set point. Particularly, the proposed FIS takes as input the internal ambient temperature of the neighborhood offices, the internal ambient temperature of the open space, and the external ambient temperature, and gives as output the new setpoint temperature of the HVAC. Each input variable takes three linguistic values ('cold', 'mild', 'warm'), which correspond to fuzzy sets that are defined by trapezoidal membership functions. The system's output variable is described by using four fuzzy sets ('cold', 'little cold', 'neutral', 'little warm', 'warm'). Also, the presented FIS uses 27 fuzzy rules of the format: "if the internal ambient temperature of the neighborhood offices is X AND the internal ambient temperature of the open space is Y and the external ambient temperature is Z, then the new setpoint temperature is W", where X, Y, Z $\in$ {('cold', 'mild', 'warm'} and Z $\in$ {'cold', 'little cold', 'neutral', 'little warm', 'warm'}.
- In [395], the authors present a decoupling fuzzy-neural network for temperature and humidity control in HVAC systems. The proposed fuzzy logic-based approach allows controllers to tune their parameters automatically.
- In [396], the authors present a fuzzy logic-based approach for fault detection with limited faulty samples to the operation of HVAC systems. Particularly, an adaptative neuro-fuzzy inference system is used to identify the existence of fault

or not. The historical data that are monitored for the building energy system are pre-processed and divided into training and testing data. This dataset is used for the ANFIS's training. Then, the real-time monitoring data, which are provided by the sensors of the building energy system are fed to the ANFIS to decide if the HVAC system's operation is normal or there is a fault.

- In [397], the authors developed a smart HVAC system, which maintain the temperature close to the occupants' thermal comfort minimizing the energy consumption. The proposed approach is based on a Mamdani-type FIS, which takes as input three environmental parameters (set point temperature, outdoor temperature, physical presence) and gives as output a multiplying factor for the set point temperature which will give the final temperature in degree Celsius. The system uses fuzzy sets with triangular membership functions to define input and output variables and a set of 18 fuzzy rules to manage the temperature of the air conditioner. The evaluation of the proposed system showed that it causes from 16 to 25% reduction of the total energy consumed by air conditioners.
- In [398], the authors propose a fuzzy logic-based approach for HVAC systems control. The aim of the proposed approach is the automatic and dynamic control of the speed of the HVAC evaporator fan, meeting the occupant's thermal preferences and comfort and resulting in a better management of energy consumption. The approach is based on a Mamdani-type FIS, which takes as input the occupant's desired set point temperature and the temperature difference between the actual space temperature and the desired temperature. The desired temperature is described through five fuzzy sets: 'very cold' (curved membership function), 'cold' (triangular membership function), 'medium' (triangular membership function), 'hot' (triangular membership function), and 'very hot' (curved membership function). The temperature difference is described, also, through five fuzzy sets: 'extra negative' (curved membership function), 'negative' (triangular membership function), 'zero error' (triangular membership function), 'positive' (triangular membership function), and 'extra positive' (curved membership function). The system's output is the speed of the evaporator's fan. It is described through six fuzzy sets: 'very fast' (curved membership function), 'fast' (triangular membership function), 'medium' (triangular membership function), 'slow' (triangular membership function), 'very slow' (curved membership function), 'off' (curved membership function).
- In [399], the authors developed an enhanced model-based predictive control practical system based on a Mamdani-type FIS. The aim of the presented fuzzy logic-based approach is to maintain the indoor thermal comfort in IoT smart space. The proposed FIS has three input parameters: (i) the predicted mean vote index, which is a widely known model for evaluating the global thermal comfort and its scale is between cold (−3) and hot (+3), (ii) the variation of the predicted mean vote index, and (iii) the outside temperature. The linguistic values for the predicted mean vote index are 'too cold' (trapezoidal membership function), 'winter comfort' (trapezoidal membership function), 'neutral' (triangular membership function), 'summer comfort' (trapezoidal membership function),

'too hot' (trapezoidal membership function). The linguistic values for variation of the predicted mean vote index are 'negative big' (trapezoidal membership function), 'negative medium' (trapezoidal membership function), 'negative small' (trapezoidal membership function), 'zero' (triangular membership function), 'positive small' (trapezoidal membership function), 'positive medium' (trapezoidal membership function), 'positive big' (trapezoidal membership function). The linguistic values for the outside temperature are 'very low' (trapezoidal membership function), 'low' (trapezoidal membership function), 'medium' (trapezoidal membership function), 'high' (trapezoidal membership function), 'very high' (trapezoidal membership function). The system's output is the fan speed, which is described by using five fuzzy sets: 'zero', 'low', 'medium', 'high', 'very high'. The membership functions of all the fuzzy sets of the output variable are trapezoidal. The presented FIS estimates dynamically the speed of the fan by applying a set of 175 fuzzy rules over the input fuzzy variables.

- In [400], the authors developed an adaptive HVAC fuzzy logic controller for better energy management in commercial facilities. The presented fuzzy logic controller is based on a Mamdani-type FIS, which takes as input the current temperature and the number of consumers and gives as output the heating and cooling functions, which need to be performed for a better energy management. Five fuzzy sets are used to describe both input variables. Particularly, the linguistic values for 'temperature' are: 'too cold', 'cold', 'warm', 'hot', 'too hot'. Also, the linguistic values for 'consumers' number' are: 'very low', 'low', 'medium', 'high', 'very high'. Both, triangular and trapezoidal membership functions are used for the definitions of the fuzzy sets of the input variables. Furthermore, both output variables are described by using four fuzzy sets, whose membership functions are triangular. These four fuzzy sets, which describe the heating and cooling functions, are: 'off', 'slow', 'medium', 'fast'.
- In [401], the authors studied the use of fuzzy logic in HVAC systems concerning the thermal comfort provision. They developed a fuzzy logic-based heater controller and a fuzzy logic-based cooler controller. Both controllers are similar. They take as input the error value, which is the difference between the actual and the desired temperature, and the error change, which is the difference between the present error value and the previous one, and give as output the incremental change in the flow regulator position. The input variables take linguistic values, which are defined through seven fuzzy sets. The proposed fuzzy logic-based systems are a zero-order Sugeno type. Therefore, the output variable takes crisp values. Furthermore, the authors developed a fuzzy logic-based window controller to allow natural cooling during moderate seasons. The fuzzy logic-based window controller is, also, a zero-order Sugeno type FIS, which takes as input the predicted mean vote index and the outdoor temperature and gives as output the window opening fraction. A simulation of the operations of the proposed fuzzy logic-based controllers showed that their application in HVAC systems reduce the dissatisfaction of occupants and the non-comfort hours.
- In [402], the authors design a smart fuzzy logic-based controller for a hybrid active/passive cooling system, which uses wind catchers for natural ventilation. The aim

of the designed smart controller is to reduce energy consumption. The proposed fuzzy system is a Mamdani-type FIS, which takes as input the temperature and airflow and gives as outputs three values of the hybrid cooling system (shutter, fan, AC). The input variable 'temperature' takes four linguistic values ('cold', 'comfortable', 'hot', 'too hot'), which are defined through Gaussian membership functions. The input variable 'airflow' takes three linguistic values ('low', 'comfortable', 'high'), which rae defined through trapezoidal membership functions. The outputs variables take the values 'ON' or 'OFF'. Therefore, the system uses 8 fuzzy rules to decide which appliance of the cooling system has to be open or close.

- In [403], the authors present a fuzzy logic-based automated controller for efficient energy management. The proposed system is based on a Mamdani-type FIS, which takes as input two fuzzy variables and gives as output two fuzzy variables. The system's input variables are: (i) the difference between the actual and the desires temperature, and (ii) the rate of temperature change. Both input variables are described through seven fuzzy sets ('large positive deviation', 'average positive deviation', 'small positive deviation', 'zero deviation', 'small negative deviation, 'average negative deviation', 'large negative deviation'), whose membership functions are triangular. The system's output variables are: (i) the operating modes of heating and cooling system, and (ii) the fan rotation speed. The output variable 'work mode' is described through six fuzzy sets ('strong cooling', 'average cooling', 'slight cooling', 'without changes, 'heating 1', 'heating 2'), whose membership functions are triangular. Also, the output variable 'fan rotation speed' is described through four fuzzy sets ('fast', 'medium', 'low', 'zero'), whose membership functions are triangular. The rule base of the system includes 49 fuzzy rules of the format 'If 'temperature difference' is X and 'range of temperature change' is Y, then 'work mode' is W and 'fan rotation speed' is G'.
- In [404], the authors present a fault diagnosis technique for HVAC systems, which is based on fuzzy logic and Bayesian belief network. It combines the benefits of knowledge-driven and data-driven diagnosis. The faults that are diagnosed by the proposed model are: CO_2 sensor fault, temperature sensor fault, damper actuator fault, heater actuator (thermostat) fault. The parameters that are described through fuzzy sets are: (i) room temperature, (ii) daily temperature, (iii) CO2 concentration, and (iv) occupancy. Three trapezoidal fuzzy sets ('low', 'middle', 'high') are used for each one of the aforementioned parameters. The combination of fuzzy logic with Bayesian belief network is used to analyze the dependencies of the sensors' signals.
- In [405], the authors use fuzzy logic to model the operation of HVAC system for energy saving and cost reduction. Particularly, the use a Mamdani-type FIS, which takes two inputs and gives one output. The input variables are: (i) error, which is defined by three fuzzy sets ('very cool', 'medium cool', 'large cool') whose membership functions are trapezoidal (ii) change of error, which is defined by three fuzzy sets ('small', 'medium', 'large') whose membership functions are trapezoidal. The output variable is the variation between the indoor air temperature and the set-point temperature, and it takes three linguistic values ('off', 'steady',

'on') which are associated with triangular fuzzy sets. The value of the FIS output is calculated by applying 8 fuzzy rules of the format "if 'error' is X and 'change of error' is Y, then 'temperature variation' is Z", where X ∈ {'very cool', 'medium cool', 'large cool'}, Y ∈ {'small', 'medium', 'large'} and Z ∈ {'off', 'steady', 'on'}. Furthermore, particle swarm optimization algorithm is used to improve the FIS outcome and find an optimal cost and power consumption.

- In [406], the authors combine fuzzy logic with machine learning to allow HVAC system to adjust automatically the indoor climate conditions in order to provide thermal comfort to building occupants. In more details, a Random Forest algorithm is used to estimate the air quality and a Mamdani-type FIS is used to adjust the temperature set-point. The designed FIS takes four inputs and gives one output. The input fuzzy variables are: (i) derivative temperature, which is described by the following fuzzy sets: 'negative' (trapezoidal membership function), 'zero' (triangular membership function), 'positive' (trapezoidal membership function); (ii) error, which is described by the following fuzzy sets: 'negative' (trapezoidal membership function), 'zero' (triangular membership function), 'positive' (trapezoidal membership function); (iii) indoor-outdoor difference, which is described by the following fuzzy sets: 'cool' (Gaussian membership function), 'same' (Gaussian membership function), 'hot' (Gaussian membership function); and (iv) humidity, which is described by the following fuzzy sets: 'small' (trapezoidal membership function), 'medium' (triangular membership function), 'high' (trapezoidal membership function). The output variable is the temperature set-point and is described by three fuzzy sets ('make cool', 'no change', 'make hot'), whose membership functions are Gaussian. For the calculation of the output nine fuzzy rules are used.
- In [407], the authors developed a Mamdani-type FIS to control more efficient the energy consumption of HVAC systems. The proposed FIS has four input variables: user presence, demand, outdoor temperature, and price. They are described through fuzzy sets, which are defined by triangular and trapezoidal membership functions. 'User presence' takes two linguistic values: 'yes' or 'no'. 'Demand' takes three linguistic values: 'low', 'medium', 'high'. 'Outdoor temperature' takes four linguistic values: 'very cold', 'cold', 'cool', 'normal'. 'Price' takes three linguistic values: 'low', 'medium', 'high'. The output fuzzy variable of the system is the setpoint temperature, which takes eight linguistic values. The fuzzy systems rule based is consisted by 72 fuzzy rules.
- In [408], the authors propose a Mamdani-type FIS to control the operation of an air conditioner maintaining the room temperature at a desired level. The presented FIS takes three parameters as input and gives as output the appropriate signal for cooling or heating. The input fuzzy variables are: (i) temperature error or difference between the current room temperature and desired value, (ii) integral error, and (iii) differential error. All input variables are described through seven fuzzy sets ('negative large', 'negative medium', 'negative small', 'zero', 'positive small', 'positive medium', 'positive large'), which are defined by Gaussian membership functions. The systems' output variable is described through five fuzzy sets ('medium cooling', 'slightly cooling', 'steady', 'slightly heating',

'medium heating'), which are defined through triangular membership function. The knowledge base of the system includes 125 fuzzy rules.

- In [409], the authors use an interval type-2 fuzzy logic system, which considers the real-time occupancy of zones in large buildings, like malls or business parks. The proposed fuzzy logic-based system is embedded in central air conditioner. It takes as input the occupancy rate change of zone and estimates the weights for energy and comfort optimization. Both input and outputs variables take linguistic values, which are defined through triangular membership functions. In addition to interval type-2 fuzzy logic system, a non-dominated sorting genetic algorithm is used to optimize the set point values for controllers to achieve the desired microclimate in each zone.
- In [410], the authors developed a fuzzy logic-based controller for thermal comfort and indoor air quality regulation in the cabin of a vehicle. The presented fuzzy logic-based controller is based on a Mamdani-type FIS, which takes as input the predicted mean vote index, the carbon dioxide concentration, and cabin temperature, and gives as output the blower velocity, the mass flow rate of the refrigerant and hot coolant, the recirculation flap, and the HVAC mode. The 'predicted mean vote index' takes the following seven linguistic values: 'cold', 'cool', 'slightly cool', 'neutral', 'slightly warm', 'warm', and 'hot'. The 'carbon dioxide concentration' takes the following four linguistic values: 'high', 'medium', 'moderate', and 'low'. The 'cabin temperature' takes the following five linguistic values: 'cold', 'cool', 'comfort', 'warm', and 'hot'. The 'blower velocity' takes the following five linguistic values: 'off', 'low', 'medium', 'high', and 'maximum'. The 'mass flow rate of the refrigerant and hot coolant' takes the following five linguistic values: 'off', 'low', 'medium', 'high', and 'maximum'. The 'recirculation flap' takes the following four linguistic values: 'low', 'moderate', 'medium', and 'high'. The 'HVAC mode' takes the following three linguistic values: 'cooling', 'off', 'heating'. The main core of the proposed FIS is the rule base, which includes 140 fuzzy rules.
- In [411], the authors developed a Mamdani-type FIS to control the ventilation system for the pharmaceutical industry. The proposed FIS takes as input the indoor temperature, the humidity, the air flow, and the microorganisms in air and gives as output the fan speed, the cold air, and the operation mode. 'Indoor temperature' is described through three fuzzy sets ('cold', 'warm', 'hot'), whose membership functions are triangular. 'Humidity' is described through three fuzzy sets ('dry', 'moderate', 'humid'), whose membership functions are Gaussian. 'Air flow' is described through three fuzzy sets ('slow', 'medium', 'fast'), whose membership functions are Gaussian. 'Microorganism' is described through three fuzzy sets ('low', 'medium', 'high'), whose membership functions are triangular. 'Fan speed' is described through five fuzzy sets ('very slow', 'slow', 'medium', 'fast', 'very fast'), whose membership functions are triangular. 'Cold air' is described through five fuzzy sets ('very slow', 'slow', 'medium', 'fast', 'very fast'), whose membership functions are triangular and trapezoidal. Finally, 'operation mode' is described through three fuzzy sets ('dehumidifier', 'normal', 'purification'), whose membership functions are trapezoidal.

Considering the above, we notice that fuzzy logic is usually used in heating and cooling systems. Also, the most-common aims of the integration of fuzzy logic into HVAC system are energy consumption reduction and occupants' thermal comfort. Furthermore, Mamdani-type FIS is the fuzzy technique that is preferred mostly to be used in HVAC controllers. In addition, the most common parameters that are adjust automatically and dynamically by the fuzzy logic-based controllers are temperature and fan speed.

References

1. E.H. Shortliife, *Computer-Based Medical Consultations: MYCIN* (Elsevier, New York, 1976). B.G. Buchanan, E.H. Shortliffe, *Rule-Based Expert Systems. MYCIN Experiments of the Stanford Heuristic Programming Project* (Addison-Wesley, Reading, MA, 1984)
2. K.P. Adlassnig, G. Kolarz, W. Scheithauser, H. Effenberger, G. Grabner, CADIAG—approaches to computer-assisted medical diagnosis. Comput. Biol. Med. **15**, 315–333 (1985)
3. C. Kulikowski, S.M. Weiss, Representation of expert knowledge for consultation: the CAS NET and EXPERT projects, in *Artificial Intelligence in Medicine*, ed. P. Szolovits (Westview Press, Boulder, CO, 1982)
4. R.A. Miller, H.E. Pople, J.D. Meyers, Internist-I, an experimental computer-based diagnostic consultant for general internal medicine. N. Engl. J. Med. **307** (1982)
5. L.A. Zadeh, Biological application of the theory of fuzzy sets and systems, in *Proc. Int. Symp. Biocybernetics of the Central Nervous System* (Little, Brown & Co., Boston, 1969), pp. 199–212
6. S. Wu, P. Mao, R. Li, Z. Cai, A.A. Heidari, J. Xia, ..., X. Chen, Evolving fuzzy k-nearest neighbors using an enhanced sine cosine algorithm: Case study of lupus nephritis. Comput. Biol. Med. **135**, 104582 (2021)
7. H. Darbandi, M. Baniasad, S. Baghdadi, A. Khandan, A. Vafaee, F. Farahmand, Automatic classification of gait patterns in children with cerebral palsy using fuzzy clustering method. Clin. Biomech. **73**, 189–194 (2020)
8. I. Valencia-Hernandez, H. Peregrina-Barreto, C.A. Reyes-Garcia, G.C. Lopez-Armas, Density map and fuzzy classification for breast density by using BI-RADS. Comput. Methods Programs Biomed. **200**, 105825 (2021)
9. A. Bárdossy, A. Blinowska, W. Kuzmicz, J. Ollitrault, M. Lewandowski, A. Przybylski, Z. Jaworski, Fuzzy logic-based diagnostic algorithm for implantable cardioverter defibrillators. Artif. Intell. Med. **60**(2), 113–121 (2014)
10. E. Ramirez, P. Melin, G. Prado-Arechiga, Hybrid model based on neural networks, type-1 and type-2 fuzzy systems for 2-lead cardiac arrhythmia classification. Expert Syst. Appl. **126**, 295–307 (2019)
11. F. Hamedan, A. Orooji, H. Sanadgol, A. Sheikhtaheri, Clinical decision support system to predict chronic kidney disease: a fuzzy expert system approach. Int. J. Med. Informatics **138**, 104134 (2020)
12. C. Ntakolia, C. Kokkotis, S. Moustakidis, D. Tsaopoulos, Identification of most important features based on a fuzzy ensemble technique: evaluation on joint space narrowing progression in knee osteoarthritis patients. Int. J. Med. Informatics **156**, 104614 (2021)
13. W.L. Zuo, Z.Y. Wang, T. Liu, H.L. Chen, Effective detection of Parkinson's disease using an adaptive fuzzy k-nearest neighbor approach. Biomed. Signal Process. Control **8**(4), 364–373 (2013)
14. R.B. Lukmanto, A. Nugroho, H. Akbar, Early detection of diabetes mellitus using feature selection and fuzzy support vector machine. Procedia Comput. Sci. **157**, 46–54 (2019)

15. P. Kora, K. Meenakshi, K. Swaraja, A. Rajani, M.K. Islam, Detection of cardiac arrhythmia using fuzzy logic. Inform. Med. Unlocked **17**, 100257 (2019)
16. B.A. Akinnuwesi, B.A. Adegbite, F. Adelowo, U. Ima-Edomwonyi, G. Fashoto, O.T. Amumeji, Decision support system for diagnosing rheumatic-musculoskeletal disease using fuzzy cognitive map technique. Inform. Med. Unlocked **18**, 100279 (2020)
17. K.R. Domínguez Hernández, A.A. Aguilar Lasserre, R. Posada Gómez, J.A. Palet Guzmán, B.E. González Sánchez, Development of an expert system as a diagnostic support of cervical cancer in atypical glandular cells, based on fuzzy logics and image interpretation. Comput. Math. Methods Med. (2013)
18. J. Singla, B. Kaur, D. Prashar, S. Jha, G.P. Joshi, K. Park, ..., C. Seo, A novel fuzzy logic-based medical expert system for diagnosis of chronic kidney disease. Mob. Inf. Syst. (2020)
19. S. Das, P.K. Ghosh, S. Kar, Hypertension diagnosis: a comparative study using fuzzy expert system and neuro fuzzy system, in *2013 IEEE International conference on fuzzy systems (FUZZ-IEEE)* (IEEE, 2013, July), pp. 1–7
20. N. Jindal, J. Singla, B. Kaur, H. Sadawarti, D. Prashar, S. Jha, ..., C. Seo, Fuzzy logic systems for diagnosis of renal cancer. Appl. Sci. **10**(10), 3464 (2020)
21. G. Casalino, R. Grassi, M. Iannotta, V. Pasquadibisceglie, G. Zaza, A hierarchical fuzzy system for risk assessment of cardiovascular disease, in *2020 IEEE Conference on Evolving and Adaptive Intelligent Systems (EAIS)* (IEEE, 2020, May), pp. 1–7
22. O. Terrada, B. Cherradi, A. Raihani, O. Bouattane (2018, December). A fuzzy medical diagnostic support system for cardiovascular diseases diagnosis using risk factors. In *2018 International Conference on Electronics, Control, Optimization and Computer Science (ICECOCS)* (pp. 1–6). IEEE.
23. H. Susanto, E. Haryatmi, Web based fuzzy expert system for lung cancer diagnosis. In *2016 2nd international conference on science in information technology (ICSITech)* (IEEE, 2016), pp. 142–146
24. A. Vashisth, G. Kaur, A. Bakshi, A, Ortho-expert: a fuzzy rule-based medical expert system for diagnosing inflammatory diseases of the knee, in *International Conference on Innovative Computing and Communications* (Springer, Singapore, 2020), pp. 77–96
25. MG Orozco-del-Castillo, RA Novelo-Cruz, J.J. Hernandez-Gomez, P.A. Mena-Zapata, E. Brito-Borges, A.E. Álvarez-Pacheco, ..., G.A. Yáñez-Casas, Fuzzy logic-based COVID-19 and other respiratory conditions pre-clinical diagnosis system, in *International Congress of Telematics and Computing* (Springer, Cham, 2020), pp. 402–419
26. A.K. Pathak, J. Arul Valan, A predictive model for heart disease diagnosis using fuzzy logic and decision tree, in *Smart Computing Paradigms: New Progresses and Challenges* (Springer, Singapore, 2020), pp. 131–140
27. O. Terrada, A. Raihani, O. Bouattane, B. Cherradi, Fuzzy cardiovascular diagnosis system using clinical data. In *2018 4th International Conference on Optimization and Applications (ICOA)* (IEEE, 2018), pp. 1–4
28. C. Reshmalakshmi, M. Sasikumar, Fuzzy inference system for osteoporosis detection, in *2016 IEEE Global Humanitarian Technology Conference (GHTC)* (IEEE, 2016), pp. 675–681
29. I. Uvaliyeva, M. Kalimoldayev, S. Rustamov, S. Belginova, Fuzzy logic for medical diagnosis of clinical and hematological symptoms, in *2019 IEEE 13th International Conference on Application of Information and Communication Technologies (AICT)* (IEEE, 2019), pp. 1–6
30. J. Kim, J. Lee, Y. Lee, Data-mining-based coronary heart disease risk prediction model using fuzzy logic and decision tree. Healthc. Inform. Res. **21**(3), 167–174 (2015)
31. W.M. Shaban, A.H. Rabie, A.I. Saleh, M.A. Abo-Elsoud, Detecting COVID-19 patients based on fuzzy inference engine and Deep Neural Network. Appl. Soft Comput. **99**, 106906 (2021)
32. M.M. Saleck, A. ElMoutaouakkil, M. Mouçouf, Tumor detection in mammography images using fuzzy C-means and GLCM texture features, in *2017 14th International Conference on Computer Graphics, Imaging and Visualization* (IEEE, 2017), pp. 122–125
33. N.F. Soliman, N.S. Ali, M.I. Aly, A.D. Algarni, W. El-Shafai, F.E. Abd El-Samie, An efficient breast cancer detection framework for medical diagnosis applications. CMC-Comput. Mater. Contin. **70**(1), 1315–1334 (2022)

34. T. Tirupal, B. Chandra Mohan, S. Srinivas Kumar, Multimodal medical image fusion based on interval-valued intuitionistic fuzzy sets, in *Machines, Mechanism and Robotics* (Springer, Singapore, 2022), pp. 965–971
35. C.L. Chowdhary, M. Mittal, P.A. Pattanaik, Z. Marszalek, An efficient segmentation and classification system in medical images using intuitionist possibilistic fuzzy C-mean clustering and fuzzy SVM algorithm. Sensors **20**(14), 3903 (2020)
36. S. Alagarsamy, Y.D. Zhang, V. Govindaraj, M.P. Rajasekaran, S. Sankaran, Smart identification of topographically variant anomalies in brain magnetic resonance imaging using a fish school-based fuzzy clustering approach. IEEE Trans. Fuzzy Syst. **29**(10), 3165–3177 (2020)
37. J.B. Monteiro-Filho, I.M.C. Albuquerque, F.L. Neto, Fish school search algorithm for constrained optimization (2017). arXiv preprint arXiv:1707.06169
38. D. Palani, K. Venkatalakshmi, An IoT based predictive modelling for predicting lung cancer using fuzzy cluster based segmentation and classification. J. Med. Syst. **43**(2), 1–12 (2019)
39. N. Otsu, A threshold selection method from gray level histograms. IEEE Trans. Syst. Man Cybern. **9**(1), 62–66 (1979)
40. S. Rajasri, D.R. Gandh, Rule based fuzzy image segmentation for the detection of breast cancer from ultrasound image, in *International Conference on Intelligent Computing and Communication for Smart World (I2C2SW)* (IEEE, 2018), pp. 254–259
41. Y.C. Joshya, P. Divyabharathi, P. Archana, N. Vikram, Automated detection of lung cancer based on neuro fuzzy technique, in *Journal of Physics: Conference Series*, vol. 1979, no. 1 (IOP Publishing, 2021), p. 012021
42. W. Cai, B. Zhai, Y. Liu, R. Liu, X. Ning, Quadratic polynomial guided fuzzy C-means and dual attention mechanism for medical image segmentation. Displays **70**, 102106 (2021)
43. R. Kumar, S. Srivastava, R. Srivastava, A fourth order PDE based fuzzy c-means approach for segmentation of microscopic biopsy images in presence of Poisson noise for cancer detection. Comput. Methods Programs Biomed. **146**, 59–68 (2017)
44. A. Selvapandian, K. Manivannan, Fusion based glioma brain tumor detection and segmentation using ANFIS classification. Comput. Methods Programs Biomed. **166**, 33–38 (2018)
45. S. Madanala, K.J. Rani, PCA-DWT based medical image fusion using non sub-sampled contourlet transform, in 2016 *International Conference on Signal Processing, Communication, Power and Embedded System (SCOPES)* (IEEE, 2016), pp. 1089–1094
46. A. Khosravanian, M. Rahmanimanesh, P. Keshavarzi, S. Mozaffari, Fast level set method for glioma brain tumor segmentation based on Superpixel fuzzy clustering and lattice Boltzmann method. Comput. Methods Programs Biomed. **198**, 105809 (2021)
47. L. Szilagyi, Z. Benyo, S.M. Szilágyi, H.S. Adam, MR brain image segmentation using an enhanced fuzzy c-means algorithm, in *Proceedings of the 25th annual international conference of the IEEE engineering in medicine and biology society (IEEE Cat. No. 03CH37439)*, vol. 1. (IEEE, 2003), pp. 724–726
48. T. Lei, X. Jia, Y. Zhang, S. Liu, H. Meng, A.K. Nandi, Superpixel-based fast fuzzy C-means clustering for color image segmentation. IEEE Trans. Fuzzy Syst. **27**(9), 1753–1766 (2018)
49. S. Balla-Arabé, X. Gao, B. Wang, A fast and robust level set method for image segmentation using fuzzy clustering and lattice Boltzmann method. IEEE Trans. Cybern. **43**(3), 910–920 (2013)
50. F.V. Farahani, A. Ahmadi, M.H.F. Zarandi, Hybrid intelligent approach for diagnosis of the lung nodule from CT images using spatial kernelized fuzzy c-means and ensemble learning. Math. Comput. Simul. **149**, 48–68 (2018)
51. S. Martinez-Muñoz, D. Ruiz-Fernandez, J.J. Galiana-Merino, Automatic abdominal aortic aneurysm segmentation in MR images. Expert Syst. Appl. **54**, 78–87 (2016)
52. A. Dixit, S. Pathak, R. Raj, C.H. Naveen, V.R. Satpute,. An efficient fuzzy based edge estimation for iris localization and pupil detection in human eye for automated cataract detection system. In 2018 9th International Conference on Computing, Communication and Networking Technologies (ICCCNT) (pp. 1–6). IEEE.
53. P. Thirumurugan, P. Shanthakumar, Brain tumor detection and diagnosis using ANFIS classifier. Int. J. Imaging Syst. Technol. **26**(2), 157–162 (2016)

54. Y. Jiang, K. Zhao, K. Xia, J. Xue, L. Zhou, Y. Ding, P. Qian, A novel distributed multitask fuzzy clustering algorithm for automatic MR brain image segmentation. J. Med. Syst. **43**(5), 1–9 (2019)
55. A. Al-Ali, O. Elharrouss, U. Qidwai, S. Al-Maaddeed, ANFIS-Net for automatic detection of COVID-19. Sci. Rep. **11**(1), 1–13 (2021)
56. P.S. Sajja, Application of fuzzy convolutional neural network for disease diagnosis: a case of Covid-19 diagnosis through CT scanned lung images, in *Tracking and Preventing Diseases with Artificial Intelligence* (Springer, Cham, 2022), pp. 177–199
57. Y. Akyuz, Effects of intelligent tutoring systems (ITS) on personalized learning (PL). Creat. Educ. **11**(6), 953–978 (2020)
58. K. Chrysafiadi, S. Papadimitriou, M. Virvou, Cognitive-based adaptive scenarios in educational games using fuzzy reasoning. Knowl.-Based Syst., 109111 (2022)
59. W.J. Clancey, R.R. Hoffman, Methods and standards for research on explainable artificial intelligence: Lessons from intel-ligent tutoring systems. Appl. AI Lett. (2021)
60. M. Virvou, E. Alepis, G.A. Tsihrintzis, L.C. Jain, Machine Learning Paradigms - Advances in Learning Analytics. Springer 2020.
61. Khazanchi, R.; Khazanchi, P. Artificial intelligence in education: A closer look into intelligent tutoring systems. In Handbook of research on critical issues in special education for school rehabilitation practices 2021, 256–277, IGI Global.
62. Chrysafiadi & Virvou, Student modeling approaches: a literature review for the last decade, Student modeling approaches: a literature review for the last decade (2013)
63. S. Papadimitriou, K. Chrysafiadi, M. Kamitsios, M. Virvou, FSP creator: a novel web service API creator of fuzzy students progress profile, in *2018 9th International Conference on Information, Intelligence, Systems and Applications (IISA)* (IEEE, 2018), pp. 1–7
64. K. Chrysafiadi, M. Virvou, Fuzzy logic in student modeling, in *Advances in Personalized Web-Based Education. Intelligent Systems Reference Library*, vol. 78 (Springer, Cham, 2015). https://doi.org/10.1007/978-3-319-12895-5_2
65. S. Sani, T.N.M. Aris, Proposal for ontology based approach to fuzzy student model design, in *2014 5th International Conference on Intelligent Systems, Modelling and Simulation* (2014), pp. 35–37, https://doi.org/10.1109/ISMS.2014.14
66. S.M. Sani, T.N.M. Aris, N.M.M.N. Sulaiman, A Fuzzy Logic approach to manage uncertainty and improve the student model prediction accuracy in design. J. Theor. Appl. Inf. Technol. **82**(3) (2015)
67. V. Bradac, B. Walek, A comprehensive adaptive system for e-learning of foreign languages. Expert Syst. Appl. **90**, 414–426 (2017)
68. C. Troussas, K. Chrysafiadi, M. Virvou, Machine learning and fuzzy logic techniques for personalized tutoring of foreign languages, in *Artificial Intelligence in Education. AIED 2018*. Lecture Notes in Computer Science, vol 10948 (Springer, Cham, 2018). https://doi.org/10.1007/978-3-319-93846-2_67
69. A. Karaci, Intelligent tutoring system model based on fuzzy logic and constraint-based student model. Neural Comput. Applic. **31**, 3619–3628 (2019). https://doi.org/10.1007/s00521-017-3311-2
70. C. Troussas, K. Chrysafiadi, M. Virvou, An intelligent adaptive fuzzy-based inference system for computer-assisted language learning. Expert. Syst. Appl. **127**, 85–96 (2019)
71. N. Singh, N.J. Ahuja, Implementation and evaluation of intelligence incorporated tutoring system. Int. J. Innov. Technol. Explor. Eng. **8**(10C) (2019)
72. M. Eryılmaz, A. Adabashi, Development of an intelligent tutoring system using Bayesian networks and fuzzy logic for a higher student academic performance. Appl. Sci. **10**(19), 6638 (2020). https://doi.org/10.3390/app10196638
73. S. Dutt, N.J. Ahuja, M. Kumar, An intelligent tutoring system architecture based on fuzzy neural network (FNN) for special education of learning disabled learners. Educ Inf Technol **27**, 2613–2633 (2022). https://doi.org/10.1007/s10639-021-10713-x
74. M. Thapliyal, N.J. Ahuja, A. Shankar et al., A differentiated learning environment in domain model for learning disabled learners. J. Comput. High. Educ. **34**, 60–82 (2022). https://doi.org/10.1007/s12528-021-09278-y

75. R.Z. Cabada, M.L.B. Estrada, Y.H. Pérez, Knowledge-based system in an affective and intelligent tutoring system, in *Current Trends on Knowledge-Based Systems. Intelligent Systems Reference Library*, ed. G. Alor-Hernández, R. Valencia-García, vol 120 (Springer, Cham, 2017). https://doi.org/10.1007/978-3-319-51905-0_5
76. K. Crockett, A. Latham, N. Whitton, On predicting learning styles in conversational intelligent tutoring systems using fuzzy decision trees. Int. J. Hum Comput Stud. **97**, 98–115 (2017)
77. L. Alfaro, C. Rivera, J. Luna-Urquizo, E. Castaneda, F. Fialho, Utilization of a neuro fuzzy model for the online detection of learning styles in adaptive e-learning systems. Int. J. Adv. Comput. Sci. Appl. **9**(12) (2018)
78. J.A. Rojas, H.E. Espitia, L.A. Bejarano, Design and optimization of a fuzzy logic system for academic performance prediction. Symmetry **13**(1), 133 (2021)
79. I. Hidayah, A.E. Permanasari, N. Ratwastuti, Student classification for academic performance prediction using neuro fuzzy in a conventional classroom, in *2013 International Conference on Information Technology And Electrical Engineering (ICITEE)* (IEEE, 2013), pp. 221–225
80. S. Maitra, S. Madan, P. Mahajan, An adaptive neural fuzzy inference system for prediction of student performance in higher education, in *2018 International Conference on Advances in Computing, Communication Control and Networking (ICACCCN)* (IEEE, 2018), pp. 1158–1163
81. T. Mansouri, A. Zare Ravasan, A. Ashrafi, A learning fuzzy cognitive map (LFCM) approach to predict student performance. J. Inf. Technol. Educ.: Res. **20**, 221–243 (2021)
82. C.S. Teh, S. Lee, M. Barawi, Predicting students' course performance based on learners' characteristics via fuzzy modelling approach. Int. J. Adv. Sci., Eng. Inf. Technol. **9**, 1944 (2019). https://doi.org/10.18517/ijaseit.9.6.10229
83. L.H. Son, H. Fujita, Neural-fuzzy with representative sets for prediction of student performance. Appl. Intell. **49**, 172–187 (2019). https://doi.org/10.1007/s10489-018-1262-7
84. P. Asopa, S. Asopa, N. Joshi, I. Mathur, Evaluating student performance using fuzzy inference system in fuzzy ITS, in *2016 International Conference on Advances in Computing, Communications and Informatics (ICACCI)*. (IEEE, 2016), pp. 1847–1851
85. S. Wen, D. Liu, Students' learning performance evaluation using a new fuzzy inference system, in *2021 2nd International Conference on Artificial Intelligence and Education (ICAIE)* (IEEE, 2021), pp. 650–656
86. A. Barlybayev, A. Sharipbay, G. Ulyukova, T. Sabyrov, B. Kuzenbayev, Student's performance evaluation by fuzzy logic. Procedia Comput. Sci. **102**, 98–105 (2016)
87. S.M. Darwish, Uncertain measurement for student performance evaluation based on selection of boosted fuzzy rules. IET Sci. Meas. Technol. **11**(2), 213–219 (2017)
88. R.S. Yadav, Application of hybrid clustering methods for student performance evaluation. Int. J. Inf. Technol. **12**(3), 749–756 (2020)
89. R. Rosadi, R. Sudrajat, B. Kharismawan, Y.A. Hambali, Student academic performance analysis using fuzzy C-means clustering, in *IOP Conference Series: Materials Science and Engineering*, vol. 166, no. 1 (IOP Publishing, 2017) p. 012036
90. N. Varela, E.S. Montero, C. Vásquez, J.G. Guiliany, C.V. Mercado, N.O. Llinas, ..., P. Palencia, Student performance assessment using clustering techniques, in *Data Mining and Big Data: 4th International Conference, DMBD 2019, Chiang Mai, Thailand, July 26–30, 2019*, Proceedings 4 (Springer, Singapore, 2019), pp. 179–188
91. A. Albayrak, M. Albayrak, Performance evaluation of practice courses using fuzzy cognitive maps, in *2016 15th International Conference on Information Technology Based Higher Education and Training (ITHET)* (IEEE, 2016), pp. 1–7
92. J.M. Babanli, Fuzzy approach for evaluation of student's performance, in *14th International Conference on Theory and Application of Fuzzy Systems and Soft Computing–ICAFS-2020*, vol. 14 (Springer International Publishing, 2021), pp. 140–147
93. A.F. Baba, F.M. Cin, D. Bakanay, A fuzzy system for evaluating students' project in engineering education. Comput. Appl. Eng. Educ. **20**(2), 287–294 (2012)
94. S. Sweta, K. Lal, Optimized fuzzy rule-based system to measure uncertainty in human decision making system, in *Soft Computing: Theories and Applications* (Springer, Singapore, 2020), pp. 799–811

95. L.A. Zadeh, Fuzzy logic = computing with words. IEEE Trans. Fuzzy Syst. **4**(2), 103–111 (1996)
96. N. Bhardwaj, P. Sharma, An advanced uncertainty measure using fuzzy soft sets: application to decision-making problems. Big Data Min. Anal. **4**(2), 94–103 (2021)
97. S. Makram, A. Mourad, G. Adnane, D. Karim, Adaptive tutoring system based on fuzzy logic. Int. J. Adv. Intell. Parad. **16**(2), 132–144 (2020)
98. M. Badaracco, L. Martínez, A fuzzy linguistic algorithm for adaptive test in Intelligent Tutoring System based on competences. Expert Syst. Appl. **40**(8), 3073–3086 (2013)
99. T. Lendyuk, S. Sachenko, S. Rippaand, G. Sapojnyk, Fuzzy rules for tests complexity changing for individual learning path construction, in *Proceedings of the 8th IEEE International Conference on Intelligent Data Acquisition and Advanced Computing Systems: Technology and Applications (IDAACS'2015)* (2015), pp. 945–948
100. K. Chrysafiadi, M. Virvou, Create dynamically adaptive test on the fly using fuzzy logic, in *2018 9th International Conference on Information, Intelligence, Systems and Applications (IISA)* (IEEE, 2018), pp. 1–8
101. K. Chrysafiadi, C. Troussas, M. Virvou, Combination of fuzzy and cognitive theories for adaptive e-assessment. Expert Syst. Appl. **161**, 113614 (2020)
102. L.W. Anderson, D.R. Krathwohl, P.W. Airasian, K.A. Cruikshank, R.E. Mayer, M.C. Pintrich Wittrock (eds.), *A taxonomy for learning, teaching and assessing. A revision of Bloom's taxonomy of educational objectives* (Addison-Wesley Longman, White Plains, NY, 2001)
103. W. Ridwan, I. Wiranto, R.D.R. Dako, Ability estimation in computerized adaptive test using Mamdani Fuzzy Inference System, in *IOP Conference Series: Materials Science and Engineering*, vol. 850, no. 1 (IOP Publishing, 2020), p. 012004
104. K. Chrysafiadi, M. Virvou, G.A. Tsihrintzis, A fuzzy-based mechanism for automatic personalized assessment in an e-learning system for computer programming. Intell. Decis. Technol., (Prepr.), 1–16 (2022)
105. Valášková, K., Klieštik, T., & Mišánková, M. (2014, January). The role of fuzzy logic in decision making process. In 2014 2nd international conference on management innovation and business innovation,Vol. 44, No. 1, pp. 143–148.
106. S.F. Ismael, A.H. Alias, A.A. Zaidan, B.B. Zaidan, H.A. Alsattar, S. Qahtan, …, R.T. Mohammed, Toward sustainable transportation: a pavement strategy selection based on the extension of dual-hesitant fuzzy multicriteria decision-making methods. IEEE Trans. Fuzzy Syst. **31**(2), 380–393 (2022)
107. T. Chen, Decision-making support for transportation and logistics combining rough set fuzzy logic algorithm, 4863–4872 (2021)
108. P. Kundu, S. Kar, M. Maiti, A fuzzy multi-criteria group decision making based on ranking interval type-2 fuzzy variables and an application to transportation mode selection problem. Soft. Comput. **21**, 3051–3062 (2017). https://doi.org/10.1007/s00500-015-1990-0
109. N. Koohathongsumrit, W. Meethom, A fuzzy decision making framework for route selection in multimodal transportation networks. Eng. Manag. J. **34**(4), 689–704 (2022). https://doi.org/10.1080/10429247.2022.2027205
110. N. Jan, J. Gwak, J. Choi, S.W. Lee, C.S. Kim, Transportation strategy decision-making process using interval-valued complex fuzzy soft information. AIMS Math. **8**(2), 3606–3633 (2023)
111. D. Pamucar, I. Gokasar, A.E. Torkayesh, M. Deveci, L. Martínez, Q. Wu, Prioritization of unmanned aerial vehicles in transportation systems using the integrated stratified fuzzy rough decision-making approach with the hamacher operator. Inf. Sci. **622**, 374–404 (2023)
112. N. Elagouz, N.C. Onat, M. Kucukvar, B. Ayvaz, A.A. Kutty, A.O. Kusakci, Integrated modelling for sustainability assessment and decision making of alternative fuel buses. Transp. Res. Part D: Transp. Environ. **117**, 103656 (2023)
113. C. Prakash, M.K. Barua, An analysis of integrated robust hybrid model for third-party reverse logistics partner selection under fuzzy environment. Resour. Conserv. Recycl. **108**, 63–81 (2016)
114. D. Kumar, Z. Rahman, F.T.S. Chan, A fuzzy AHP and fuzzy multi objective linear programming model for order allocation in a sustainable supply chain a case study. Int. J. Comput. Integr. Manuf. **30**, 535–551 (2017)

115. K. Zimmer, M. Fröhling, P. Breun, F. Schultmann, Assessing social risks of global supply chains: a quantitative analytical approach and its application to supplier selection in the German automotive industry. J. Clean. Prod. **149**, 96–109 (2017)
116. B.O.V. Reyes, J.C. Colmenero, A fuzzy decision-making methodology for third-party logistics selection in the existence of the company's policy priorities: case of household appliance manufacturer. Int. J. Logist. Syst. Manag. **43**(2), 193–237 (2022)
117. J. Song, L. Jiang, Z. Liu, X. Leng, Z. He, Selection of third-party reverse logistics service provider based on intuitionistic fuzzy multi-criteria decision making. Systems **10**(5), 188 (2022)
118. C.N. Wang, N.A.T. Nguyen, T.T. Dang, C.M. Lu, A compromised decision-making approach to third-party logistics selection in sustainable supply chain using fuzzy AHP and fuzzy VIKOR methods. Mathematics **9**(8), 886 (2021)
119. S. Jovčić, P. Průša, M. Dobrodolac, L. Švadlenka, A proposal for a decision-making tool in third-party logistics (3PL) provider selection based on multi-criteria analysis and the fuzzy approach. Sustainability **11**(15), 4236 (2019)
120. R. Maharjan, S. Hanaoka, Fuzzy multi-attribute group decision making to identify the order of establishing temporary logistics hubs during disaster response. J. Human. Logist. Supply Chain. Manag. (2019)
121. Ocampo, L. A., Himang, C. M., Kumar, A., & Brezocnik, M. (2019). A novel multiple criteria decision-making approach based on fuzzy DEMATEL, fuzzy ANP and fuzzy AHP for mapping collection and distribution centers in reverse logistics. Adv. Prod. Eng. Manag. **14**(3)
122. C. Prakash, M.K. Barua, A multi-criteria decision-making approach for prioritizing reverse logistics adoption barriers under fuzzy environment: case of Indian electronics industry. Glob. Bus. Rev. **17**(5), 1107–1124 (2016)
123. J. Wang, J.Q. Wang, Z.P. Tian, D.Y. Zhao, A multihesitant fuzzy linguistic multicriteria decision-making approach for logistics outsourcing with incomplete weight information. Int. Trans. Oper. Res. **25**(3), 831–856 (2018)
124. A. Calabrese, R. Costa, N. Levialdi, T. Menichini, Integrating sustainability into strategic decision-making: a fuzzy AHP method for the selection of relevant sustainability issues. Technol. Forecast. Soc. Chang. **139**, 155–168 (2019)
125. S. Narayanamoorthy, J.V. Brainy, R.A. Shalwala, T.R. Alsenani, A. Ahmadian, D. Kang, An enhanced fuzzy decision making approach for the assessment of sustainable energy storage systems. Sustain. Energy, Grids Netw. **33**, 100962 (2023)
126. A. Canan, B. Ahmet, T.G. Tekin, Sustainability analysis of different hydrogen production options using hesitant fuzzy AHP. Int. J. Hydrogen Energy **43**, 18059–18076 (2018)
127. B. Wang, J. Song, J. Ren, K. Li, H. Duan, X.E. Wang, Selecting sustainable energy conversion technologies for agricultural residues: a fuzzy AHP-VIKOR based prioritization from life cycle perspective. Resour. Conserv. Recycl. **142**, 78–87 (2019)
128. A. Erjaee, S. Hendiani, S. Moradi et al., Healthcare sustainability evaluation using a hybrid fuzzy multi-criteria decision-making model. Int. J. Fuzzy Syst. **24**, 1182–1202 (2022). https://doi.org/10.1007/s40815-021-01128-4
129. A. Yildizbasi, Y. Arioz, Green supplier selection in new era for sustainability: a novel method for integrating big data analytics and a hybrid fuzzy multi-criteria decision making. Soft Comput. **26**, 253–270 (2022). https://doi.org/10.1007/s00500-021-06477-8
130. C. Tang, D. Xu, N. Chen, Sustainability prioritization of sewage sludge to energy scenarios with hybrid-data consideration: a fuzzy decision-making framework based on full consistency method and fusion ranking model. Environ. Sci. Pollut. Res. **28**, 5548–5565 (2021). https://doi.org/10.1007/s11356-020-10544-2
131. E. Yadegaridehkordi, M. Hourmand, M. Nilashi, E. Alsolami, S. Samad, M. Mahmoud, …, L. Shuib, Assessment of sustainability indicators for green building manufacturing using fuzzy multi-criteria decision making approach. J. Clean. Prod. **277**, 122905 (2020)
132. J. Andria, G. di Tollo, R. Pesenti, Fuzzy multi-criteria decision-making: an entropy-based approach to assess tourism sustainability. Tour. Econ. **27**(1), 168–186 (2021)

133. A. Fetanat, M. Tayebi, H. Mofid, Combining life cycle sustainability assessment and fuzzy multicriteria decision making method for prioritizing the flare technologies in the oil, gas, and chemical plants. Environ. Prog. Sustain. Energy **41**(5), e13837 (2022)
134. S.D. Attri, S. Singh, A. Dhar, S. Powar, Multi-attribute sustainability assessment of wastewater treatment technologies using combined fuzzy multi-criteria decision-making (2022)
135. J. Ren, H. Liang, Measuring the sustainability of marine fuels: a fuzzy group multi-criteria decision making approach. Transp. Res. Part D: Transp. Environ. **54**, 12–29 (2017)
136. N. Banaeian, H. Mobli, B. Fahimnia, I.E. Nielsen, M. Omid, Green supplier selection using fuzzy group decision making methods: a case study from the agri-food industry. Comput. Oper. Res. **89**, 337–347 (2018)
137. D. Kannan, R. Khodaverdi, L. Olfat, A. Jafarian, A. Diabat, Integrated fuzzy multi criteria decision making method and multi-objective programming approach for supplier selection and order allocation in a green supply chain. J. Clean. Prod. **47**, 355–367 (2013)
138. A.K. Kar, Revisiting the supplier selection problem: An integrated approach for group decision support. Expert Syst. Appl. **41**, 2762–2771 (2014)
139. H.M. Wang Chen, S.-Y. Chou, Q.D. Luu, T.H.-K. Yu, A fuzzy MCDM approach for green supplier selection from the economic and environmental aspects. Math. Probl. Eng. **2016**, 1–10 (2016)
140. I. Sultana, I. Ahmed, A. Azeem, An integrated approach for multiple criteria supplier selection combining Fuzzy Delphi, Fuzzy AHP & Fuzzy TOPSIS. J. Intell. Fuzzy Syst. **29**, 1273–1287 (2015)
141. A.U. Rahman, M. Saeed, M.A. Mohammed, A. Majumdar, O. Thinnukool, Supplier selection through multicriteria decision-making algorithmic approach based on rough approximation of Fuzzy hypersoft sets for construction project. Buildings **12**(7), 940 (2022)
142. Y. Xing, M. Cao, Y. Liu, M. Zhou, J. Wu, A Choquet integral based interval Type-2 trapezoidal fuzzy multiple attribute group decision making for Sustainable Supplier Selection. Comput. Ind. Eng. **165**, 107935 (2022)
143. V.T. Nguyen, N.H. Hai, N.T.K. Lan, Spherical fuzzy multicriteria decision-making model for wind turbine supplier selection in a renewable energy project. Energies **15**(3), 713 (2022)
144. M.M. Islam, M. Arakawa, Hesitant fuzzy-based integrated multi-criteria group decision-making model for supplier selection. J. Adv. Mech. Des., Syst., Manuf. **16**(4), JAMDSM0034 (2022)
145. D. Pamucar, A.E. Torkayesh, S. Biswas, Supplier selection in healthcare supply chain management during the COVID-19 pandemic: a novel fuzzy rough decision-making approach. Ann. Oper. Res., 1–43 (2022)
146. N.B.T. Nguyen, G.H. Lin, T.T. Dang, A two phase integrated fuzzy decision-making framework for green supplier selection in the coffee bean supply chain. Mathematics **9**(16), 1923 (2021)
147. Y. Dong, et al., A Novel stochastic group decision-making framework with dual hesitant fuzzy soft set for resilient supplier selection', 1049–1067 (2021)
148. S. Abdullah, Multicriteria group decision-making for supplier selection based on intuitionistic cubic fuzzy aggregation operators. Int. J. Fuzzy Syst. **22**, 810–823 (2020)
149. F. Zhou, T.Y. Chen, An integrated multicriteria group decision-making approach for green supplier selection under Pythagorean fuzzy scenarios. IEEE Access **8**, 165216–165231 (2020)
150. T. Gegovska, R. Köker, T. Çakar, Green supplier selection using fuzzy multiple-criteria decision-making methods and artificial neural networks. Comput. Intell. Neurosci. (2020)
151. C.N. Wang, V.T. Nguyen, J.T. Chyou, T.F. Lin, T.N. Nguyen, Fuzzy multicriteria decision-making model (MCDM) for raw materials supplier selection in plastics industry. Mathematics **7**(10), 981 (2019)
152. A.K. Bera, D.K. Jana, D. Banerjee, T. Nandy, Multiple-criteria fuzzy group decision-making with multi-choice goal programming for supplier selection: a case study. Discret. Math., Algorithms Appl. **11**(03), 1950029 (2019)
153. C. Yu, Y. Shao, K. Wang, L. Zhang, A group decision making sustainable supplier selection approach using extended TOPSIS under interval-valued Pythagorean fuzzy environment. Expert Syst. Appl. **121**, 1–17 (2019)

154. S.I. Mari, M.S. Memon, M.B. Ramzan, S.M. Qureshi, M.W. Iqbal, Interactive fuzzy multi criteria decision making approach for supplier selection and order allocation in a resilient supply chain. Mathematics **7**(2), 137 (2019)
155. X. Tang, G. Wei, H. Gao, Models for multiple attribute decision making with interval-valued pythagorean fuzzy muirhead mean operators and their application to green suppliers selection. Informatica **30**(1), 153–186 (2019)
156. L.A. Puška, S. Kozarević, Ž. Stević, J. Stovrag, A new way of applying interval fuzzy logic in group decision making for supplier selection. Econ. Comput. Econ. Cybern. Stud. Res. **52**(2) (2018)
157. G. Secundo, D. Magarielli, E. Esposito, G. Passiante, Supporting decision-making in service supplier selection using a hybrid fuzzy extended AHP approach: a case study. Bus. Process. Manag. J. (2017)
158. P. Bhatia, N. Diaz-Elsayed, Facilitating decision-making for the adoption of smart manufacturing technologies by SMEs via fuzzy TOPSIS. Int. J. Prod. Econ., 108762 (2023)
159. Y.T. Leong, R.T. Raymond, B.A. Kathleen, I.M.L. Chew, Fuzzy analytic hierarchy process and targeting for inter-plant chilled and cooling water net- work synthesis. J. Clean. Prod. **110**, 40–53 (2016)
160. R. Parameshwaran, S.P. Kumar, K. Saravanakumar, An integrated fuzzy mcdm based approach for robot selection considering objective and subjective criteria. Appl. Soft Comput. **26**, 31–41 (2015)
161. R. Zare, J. Nouri, M.A. Abdoli, F. Atabi, M. Alavi, The integrated fuzzy ahp and goal programing model based on lca results for industrial waste management by using the nearest weighted approximation of fns: Aluminum industry arak-iran. Adv. Mater. Sci. Eng., 1–13 (2016)
162. X Luo, S. Guo, B. Du, J. Guo, P. Jiang, T. Tan, Multi-criteria decision-making of manufacturing resources allocation for complex product system based on intuitionistic fuzzy information entropy and TOPSIS. Complex Intell. Syst., 1–20 (2023)
163. A. Saha, D. Pamucar, O.F. Gorcun, A.R. Mishra, Warehouse site selection for the automotive industry using a fermatean fuzzy-based decision-making approach. Expert Syst. Appl. **211**, 118497 (2023)
164. P. Ponce, C. Pérez, A.R. Fayek, A. Molina, Solar Energy implementation in manufacturing industry using multi-criteria decision-making fuzzy TOPSIS and S4 framework. Energies **15**(23), 8838 (2022)
165. Y. S. Ozdemir, A spherical fuzzy multi-criteria decision-making model for industry 4.0 performance measurement. Axioms **11**(7), 325 (2022)
166. H. Reda, A. Dvivedi, Decision-making on the selection of lean tools using fuzzy QFD and FMEA approach in the manufacturing industry. Expert Syst. Appl. **192**, 116416 (2022)
167. S.K. Paul, P. Chowdhury, K. Ahsan, S.M. Ali, G. Kabir, An advanced decision-making model for evaluating manufacturing plant locations using fuzzy inference system. Expert Syst. Appl. **191**, 116378 (2022)
168. S. Vahabi Nejat, S. Avakh Darestani, M. Omidvari, M.A. Adibi, Evaluation of green lean production in textile industry: a hybrid fuzzy decision-making framework. Environ. Sci. Pollut. Res., 1–22 (2021)
169. K.S. Chen, C.M. Yu, Fuzzy decision-making model for process quality improvement of machine tool industry chain. J. Intell. Fuzzy Syst. **42**(3), 1547–1558 (2022)
170. F.K.A. Alazemi, M.K.A.B.M. Ariffin, F.B. Mustapha, E.E.B. Supeni, A comprehensive fuzzy decision-making method for minimizing completion time in manufacturing process in supply chains. Mathematics **9**(22), 2919 (2021)
171. J.H. Dahooie, N. Mohammadi, T. Daim, A.S. Vanaki, E.K. Zavadskas, Matching of technological forecasting technique to a technology using fuzzy multi-attribute decision-making methods: Case study from the aerospace industry. Technol. Soc. **67**, 101707 (2021)
172. R. Mishra, A hybrid multi-criteria decision-making approach to assess the enablers of manufacturing flexibility under fuzzy environment. Int. J. Qual. Reliab. Manag. **38**(4), 955–985 (2021)

173. C.N. Wang, V.T. Nguyen, J.C. Kao, C.C. Chen, V.T. Nguyen, Multi-criteria decision-making methods in fuzzy decision problems: a case study in the frozen shrimp industry. Symmetry **13**(3), 370 (2021)
174. Y. Yan, D. Chu, Evaluation of enterprise management innovation in manufacturing industry using fuzzy multicriteria decision-making under the background of big data. Math. Probl. Eng. **2021**, 1–10 (2021)
175. M. Bhatia, S.K. Sood, R. Kumari, Fuzzy-inspired decision making for dependability recommendation in e-commerce industry. Intell. Decis. Technol. **14**(2), 181–197 (2020)
176. T.O. Sangodapo, B.O. Onasanya, S. Mayerova-Hoskova, Decision-making with fuzzy soft matrix using a revised method: a case of medical diagnosis of diseases. Mathematics **9**(18), 2327 (2021)
177. A. Mardani, R.E. Hooker, S. Ozkul, S. Yifan, M. Nilashi, H.Z. Sabzi, G.C. Fei, Application of decision making and fuzzy sets theory to evaluate the healthcare and medical problems: a review of three decades of research with recent developments. Expert Syst. Appl. **137**, 202–231 (2019)
178. V.F. Reyna, S. Edelson, B. Hayes, D. Garavito, Supporting health and medical decision making: findings and insights from fuzzy-trace theory. Med. Decis. Making **42**(6), 741–754 (2022)
179. A. Hein, L.J. Meier, A.M. Buyx, K. Diepold, A fuzzy-cognitive-maps approach to decision-making in medical ethics, in *2022 IEEE International Conference on Fuzzy Systems (FUZZ-IEEE), Padua, Italy* (2022), pp. 1–8, https://doi.org/10.1109/FUZZ-IEEE55066.2022.9882615.
180. P. Dutta, Medical decision making using generalized interval-valued fuzzy numbers. New Math. Nat. Comput. **17**(02), 439–479 (2021)
181. P. Dutta, D. Doley, Fuzzy decision making for medical diagnosis using arithmetic of generalised parabolic fuzzy numbers. Granul. Comput. **6**, 377–388 (2021)
182. M. Kirişci, A case study for medical decision making with the fuzzy soft sets. Afr. Mat. **31**(3–4), 557–564 (2020)
183. M.I. Tariq, N.A. Mian, A. Sohail, T. Alyas, R. Ahmad, Evaluation of the challenges in the Internet of medical things with multicriteria decision making (AHP and TOPSIS) to overcome its obstruction under fuzzy environment. Mob. Inf. Syst. **2020**, 1–19 (2020)
184. A. Guleria, R.K. Bajaj, On Pythagorean fuzzy soft matrices, operations and their applications in decision making and medical diagnosis. Soft. Comput. **23**, 7889–7900 (2019)
185. R.S. Thakur, Intelligent decision making in medical data using association rules mining and fuzzy analytic hierarchy process. Int. J. Recent Technol. Eng. (IJRTE) **7**(6), 22777–23878 (2019)
186. P. Dutta, S. Goala, Fuzzy decision making in medical diagnosis using an advanced distance measure on intuitionistic fuzzy sets. Open Cybern. Syst. J. 12(1) (2018)
187. F. Xiao, A hybrid fuzzy soft sets decision making method in medical diagnosis. IEEE Access **6**, 25300–25312 (2018). https://doi.org/10.1109/ACCESS.2018.2820099
188. P. Dutta, S.R. Dash, Medical decision making via the arithmetic of generalized triangular fuzzy numbers. Open Cybern. Syst. J. 12(1) (2018)
189. L. Wang, J.J. Peng, J.Q. Wang, A multi-criteria decision-making framework for risk ranking of energy performance contracting project under picture fuzzy environment. J. Clean. Prod. **191**, 105–118 (2018)
190. S.M. Peng, Study on enterprise risk management assessment based on picture fuzzy multiple attribute decision-making method. J. Intell. Fuzzy Syst. **33**(6), 3451–3458 (2017)
191. D. Zhu, Z. Li, A.R. Mishra, Evaluation of the critical success factors of dynamic enterprise risk management in manufacturing SMEs using an integrated fuzzy decision-making model. Technol. Forecast. Soc. Chang. **186**, 122137 (2023)
192. S. Shojaeimehr, D. Rahmani, Risk management of photovoltaic power plants using a novel fuzzy multi-criteria decision-making method based on prospect theory: a sustainable development approach. Energy Convers. Manag.: X **16**, 100293 (2022)

193. W. Hussain, M.R. Raza, M.A. Jan, J.M. Merigó, H. Gao, Cloud risk management with OWA-LSTM and fuzzy linguistic decision making. IEEE Trans. Fuzzy Syst. **30**(11), 4657–4666 (2022). https://doi.org/10.1109/TFUZZ.2022.3157951
194. M. Orojloo, S.M.H. Shahdany, A. Roozbahani, Developing an integrated risk management framework for agricultural water conveyance and distribution systems within fuzzy decision making approaches. Sci. Total Environ. **627**, 1363–1376 (2018)
195. A. Azarnivand, A. Malekian, Analysis of flood risk management strategies based on a group decision making process via interval-valued intuitionistic fuzzy numbers. Water Resour. Manage **30**, 1903–1921 (2016)
196. M. Karimi, F. Moztarzadeh, A. Pakzad, A. Beynaghi and M. Mozafari, Application of Fuzzy TOPSIS for group decision making in evaluating financial risk management, in *2012 International Conference on Innovation Management and Technology Research, Malacca, Malaysia* (2012), pp. 215–219. https://doi.org/10.1109/ICIMTR.2012.6236391.
197. B. Efe, An integrated fuzzy multi criteria group decision making approach for ERP system selection. Appl. Soft Comput. **38**, 106–117 (2016)
198. V. Arya, S. Kumar, Multi-criteria decision making problem for evaluating ERP system using entropy weighting approach and q-rung orthopair fuzzy TODIM. Granul. Comput. **6**(4), 977–989 (2021)
199. A. Kazemi, S.R. Saeidi, M. Azizmohammadi, Selecting an ERP system using multi-criteria decision making method: a goal programming and fuzzy approach. Int. J. Bus. Inf. Syst. **16**(1), 55–71 (2014)
200. Amalnick M., Ansarinejad A., Ansarinejad S., Hatami-Shirkouhi L., A group decision making approach for evaluation of ERP critical success factors using fuzzy AHP, in *2010 Fourth UKSim European Symposium on Computer Modeling and Simulation, Pisa, Italy* (2010), pp. 212-217. https://doi.org/10.1109/EMS.2010.42
201. C. Kahraman, A. Beskese, I. Kaya, Selection among ERP outsourcing alternatives using a fuzzy multi-criteria decision making methodology. Int. J. Prod. Res. **48**(2), 547–566 (2010)
202. W. Zhou, Z. Xu, W. Zhou, Z. Xu, Investment decision making based on the hesitant fuzzy preference envelopment analysis. Qual. Invest. Decis.-Mak. Methods Hesitant Fuzzy Environ., 75–94 (2020)
203. A. Borovička, New fuzzy multiple criteria evaluation method as a support for investment decision making under uncertainty, in *Scientific papers of the University of Pardubice. Series D, Faculty of Economics and Administration*. 47/2019 (2019)
204. H. Dinçer, S. Yüksel, Selecting investment strategies for European Tourism Industry using the hybrid decision making approach based on interval type-2 fuzzy sets. J. Intell. Fuzzy Syst. **37**(1), 1343–1356 (2019)
205. Z. Li, D. Sun, S. Zeng, Intuitionistic fuzzy multiple attribute decision-making model based on weighted induced distance measure and its application to investment selection. Symmetry **10**(7), 261 (2018)
206. Kozlova, M., Collan, M., & Luukka, P. (2018). New investment decision-making tool that combines a fuzzy inference system with real option analysis.,
207. A. Mukherjee and A. K. Das, "Application of Interval Valued Intuitionistic Fuzzy Soft Set in Investment Decision Making," 2015 Fifth International Conference on Advances in Computing and Communications (ICACC), Kochi, India, 2015, pp. 61–64, doi: https://doi.org/10.1109/ICACC.2015.37.
208. S. Zeng, Y. Xiao, TOPSIS method for intuitionistic fuzzy multiple-criteria decision making and its application to investment selection. Kybernetes (2016)
209. M. Sachdeva, R. Lehal, S. Gupta, S. Gupta, Influence of contextual factors on investment decision-making: a fuzzy-AHP approach. J. Asia Bus. Stud. (2022)
210. H. Dinçer, T. Aksoy, S. Yüksel, U. Hacioglu, Golden cut-oriented q-rung orthopair fuzzy decision-making approach to evaluation of renewable energy alternatives for microgeneration system investments. Math. Probl. Eng. (2022)
211. Y. Wu, T. Zhang, L. Yi, Regional energy internet project investment decision making framework through interval type-2 fuzzy number based Choquet integral fuzzy synthetic model. Appl. Soft Comput. **111**, 107718 (2021)

212. H. Jiang, B.Q. Hu, A novel three-way group investment decision model under intuitionistic fuzzy multi-attribute group decision-making environment. Inf. Sci. **569**, 557–581 (2021)
213. X. Liu, Z. Wang, S. Zhang, Y. Chen, Investment decision making along the B&R using critic approach in probabilistic hesitant fuzzy environment. J. Bus. Econ. Manag. **21**(6), 1683–1706 (2020)
214. D. Luo, S. Zeng, G. Yu, Pythagorean fuzzy investment multiple attribute decision making method based on combined aggregation method. J. Intell. Fuzzy Syst. **39**(1), 949–959 (2020)
215. Y. Han, R. Yan, C. Gou, An integrated multiple attribute decision making methodology for quality evaluation of innovation and entrepreneurship education with interval-valued intuitionistic fuzzy information. J. Intell. & Fuzzy Syst., (Prepr.), 1–19
216. I. Tadić, B. Marasović, I. Jerković, Fuzzy multicriteria model to support decision making during the selection process of teaching and research staff in higher education. Syst. Res. Behav. Sci. **39**(4), 867–885 (2022)
217. J. Wang, Y. Zhou, Multi-attribute group decision-making based on interval-valued q-Rung Orthopair fuzzy power generalized Maclaurin symmetric mean operator and its application in online education platform performance evaluation. Information **12**(9), 372 (2021)
218. H.B. Liu, Y. Liu, L. Xu, S. Abdullah, Multi-attribute group decision-making for online education live platform selection based on linguistic intuitionistic cubic fuzzy aggregation operators. Comput. Appl. Math. **40**, 1–34 (2021)
219. Z. Liao, H. Liao, M. Tang, A. Al-Barakati, C. Llopis-Albert, A Choquet integral-based hesitant fuzzy gained and lost dominance score method for multi-criteria group decision making considering the risk preferences of experts: case study of higher business education evaluation. Inf. Fusion **62**, 121–133 (2020)
220. P. Koltharkar, K.K. Eldhose, R. Sridharan, Application of fuzzy TOPSIS for the prioritization of students' requirements in higher education institutions: a case study: a multi-criteria decision making approach, in *2020 International Conference on System, Computation, Automation and Networking (ICSCAN), Pondicherry, India* (2020), pp. 1–7. https://doi.org/10.1109/ICSCAN49426.2020.9262329.
221. K.H. Tong, Q.L.H.T.T. Nguyen, T.T.M. Nguyen, P.T. Nguyen, N.B. Vu, Applying the fuzzy decision-making method for program evaluation and management policy of Vietnamese higher education J. Asian Financ., Econ. Bus. **7**(9)719–726 (2020)
222. D. Shukla, Modeling systems thinking in action among higher education leaders with fuzzy multi-criteria decision making. Manag. Marketing. Chall.S Knowl. Soc. **13**(2), 946–965 (2018)
223. K. Ravichandran, K. Venkatesh, K. Saravanan, R. Muruganandham, S. Muralidharan, Applying fuzzy logic decision-making approach to analyse the students' self-management perspective in professional education. Int. J. Enterp. Netw. Manag. **7**(4), 285–297 (2016)
224. A. Peña-Ayala, J.H. Sossa-Azuela, Decision making by rule-based fuzzy cognitive maps: an approach to implement student-centered education. Fuzzy cognitive maps for applied sciences and engineering: From fundamentals to extensions and learning algorithms (2014), pp. 107–120
225. N. Demirtas, S. Özgürler, M. Özgürler, A.F. Güneri, Selecting e-purse smart card technology via fuzzy ahp and anp. J. Appl. Math. (2014)
226. Ž Stević, E.K. Zavadskas, F.M. Tawfiq, F. Tchier, T. Davidov, Fuzzy multicriteria decision-making model based on z numbers for the evaluation of information technology for order picking in warehouses. Appl. Sci. **12**(24), 12533 (2022)
227. D. Abdul, J. Wenqi, Evaluating appropriate communication technology for smart grid by using a comprehensive decision-making approach fuzzy TOPSIS. Soft. Comput. **26**(17), 8521–8536 (2022)
228. L. Maretto, M. Faccio, D. Battini, A multi-criteria decision-making model based on fuzzy logic and AHP for the selection of digital technologies. IFAC-PapersOnLine **55**(2), 319–324 (2022)
229. X. Chen, C. Hao, Research on Intelligent Matching Technology of Marine Electric Propulsion System Based on Fuzzy Multiple Decision Making, in *IOP Conference Series: Earth and Environmental Science*, vol. 585, no. 1 (IOP Publishing, 2020), p. 012118

230. J.H. Dahooie, A.S. Vanaki, N. Mohammadi, Choosing the appropriate system for cloud computing implementation by using the interval-valued intuitionistic fuzzy CODAS multi-attribute decision-making method (case study: faculty of new sciences and technologies of Tehran University). IEEE Trans. Eng. Manage. **67**(3), 855–868 (2020). https://doi.org/10.1109/TEM.2018.2884866
231. S. Domínguez, M.C. Carnero, Fuzzy multicriteria modelling of decision making in the renewal of healthcare technologies. Mathematics **8**(6), 944 (2020)
232. X. Tang, G. Wei, H. Gao, Pythagorean fuzzy Muirhead mean operators in multiple attribute decision making for evaluating of emerging technology commercialization. Economic research-Ekonomska istraživanja **32**(1), 1667–1696 (2019)
233. B. Oztaysi, S. CevikOnar, S. Seker, C. Kahraman, Water treatment technology selection using hesitant Pythagorean fuzzy hierachical decision making. J. Intell. Fuzzy Syst. **37**(1), 867–884 (2019)
234. P. Lu, X. Yang, Z.J. Wang, Fuzzy group consensus decision making and its use in selecting energy-saving and low-carbon technology schemes in star hotels. Int. J. Environ. Res. Public Health **15**(9), 2057 (2018)
235. O. Dogan, B. Öztaysi, In-store behavioral analytics technology selection using fuzzy decision making. J. Enterp. Inf. Manag. (2018)
236. M.A.B. Promentilla, J.I.B. Janairo, D.E.C. Yu, C.M.J. Pausta, A.B. Beltran, A.P. Huelgas-Orbecido, … , R.R. Tan, A stochastic fuzzy multi-criteria decision-making model for optimal selection of clean technologies. J. Clean. Prod. **183**, 1289–1299 (2018)
237. G. Büyüközkan, F. Göçer and O. Feyzioğlu, Cloud computing technology selection based on interval valued intuitionistic fuzzy group decision making using MULTIMOORA approach, in *2017 IEEE International Conference on Fuzzy Systems (FUZZ-IEEE), Naples, Italy* (2017), pp. 1–6, https://doi.org/10.1109/FUZZ-IEEE.2017.8015682
238. T.T. Chen, T.C. Su, Fuzzy-based decision-making applied to performance evaluation in value engineering. J. Chin. Inst. Eng. **40**(3), 200–206 (2017)
239. M. Mahjouri, M.B. Ishak, A. Torabian, L. Abd Manaf, N. Halimoon, J. Ghoddusi, Optimal selection of Iron and Steel wastewater treatment technology using integrated multi-criteria decision-making techniques and fuzzy logic. Process Saf. Environ. Prot. **107**, 54–68 (2017)
240. R.R. Yager, Pythagorean fuzzy subsets, in *Proc Joint IFSA World Congress and NAFIPS, Annual Meeting, Edmonton, Canada; June 24–28* (2013), pp. 57–61. https://doi.org/10.1109/IFSA-NAFIPS.2013.6608375
241. R.R. Yager, A.M. Abbasov, Pythagorean membership grades, complex numbers, and decision making: pythagorean membership grades and fuzzy subsets. Int. J. Intell. Syst. **28**, 436–452 (2013)
242. K. Atanassov, *Intuitionistic Fuzzy Sets: Theory and Applications* (Springer, Heidelberg, 1999)
243. R.R. Yager, Generalized orthopair fuzzy sets. IEEE Trans Fuzzy Syst. **25**(5), 1222–1230 (2017)
244. H. Garg (ed.), *q-Rung Orthopair Fuzzy Sets: Theory and Applications* (Springer Nature, 2022)
245. G. Choquet, Theory of capacities. Annales de l'Institut Fourier. **5**, 131–295 (1953). https://doi.org/10.5802/aif.5
246. M. Grabisch, The application of fuzzy integrals in multicriteria decision making. Eur. J. Oper. Res. **89**(3), 445–456 (1996). https://doi.org/10.1016/0377-2217(95)00176-X
247. R.M. Rodriguez, L. Martinez, F. Herrera, Hesitant fuzzy linguistic term sets for decision making. IEEE Trans. Fuzzy Syst. **20**(1), 109–119 (2012). https://doi.org/10.1109/TFUZZ.2011.2170076
248. T. Saaty, J. Alexander, *Conflict Resolution: The Analytic Hierarchy Process* (Praeger, New York, New York, 1989)
249. Y. Liu, C.M. Eckert, C. Earl, A review of fuzzy AHP methods for decision-making with subjective judgements. Expert. Syst. Appl. **161**, 113738 (2020)
250. W.R.W. Mohd, L. Abdullah, Aggregation methods in group decision making: a decade survey. Informatica **41**(1) (2017)

251. J. Krejčí, J. Stoklasa, Aggregation in the analytic hierarchy process: Why weighted geometric mean should be used instead of weighted arithmetic mean. Expert Syst. Appl. **114**, 97–106 (2018)
252. M.J. Campión, R.G. Catalan, E. Induráin, I. Lizasoain, A. Raventós-Pujol, Ó. Valero, Geometrical aggregation of finite fuzzy sets. Int. J. Approx. Reason. **103**, 248–266 (2018)
253. J. Fodor, Aggregation functions in fuzzy systems. Asp. Soft Comput., Intell. Robot. Control. **241**, 25 (2009)
254. J. Vaníček, I. Vrana, S. Aly, Fuzzy aggregation and averaging for group decision making: a generalization and survey. Knowl.-Based Syst. **22**(1), 79–84 (2009)
255. R.R. Yager, Aggregation operators and fuzzy systems modeling. Fuzzy Sets Syst. **67**(2), 129–145 (1994)
256. S.-M. Chen, Aggregating fuzzy opinions in the group decision-making environment. J. Cybern. **29**, 363–376 (1998)
257. S. Dai, L. Bi, B. Hu, Interval-valued complex fuzzy geometric aggregation operators and their application to decision making. Math. Probl. Eng. (2020)
258. Z. Xu, J. Chen, On geometric aggregation over interval-valued intuitionistic fuzzy information, in *Fourth International Conference on Fuzzy Systems and Knowledge Discovery (FSKD 2007)*, vol. 2. (IEEE, 2007), pp. 466–471 pp. 466–471
259. P. Liu, Some geometric aggregation operators based on interval intuitionistic uncertain linguistic variables and their application to group decision making. Appl. Math. Model. **37**(4), 2430–2444 (2013)
260. Z. Xu, Intuitionistic preference relations and their application in group decision making. Inf. Sci. Int. J. **177**, 2363–2379 (2007)
261. A. Görener, B. Ayvaz, A.O. Ku"sakcı, E. Altınok, A hybrid type-2 fuzzy based supplier performance evaluation methodology: The Turkish Airlines technic case Appl. Soft Comput. **56**, 436–445 (2017)
262. J. Rezaei, R. Ortt, Multi-criteria supplier segmentation using a fuzzy preference relations based AHP. Eur. J. Oper. Res. **225**, 75–84 (2013)
263. P.J.M. van Laarhoven, W. Pedrycz, A fuzzy extension of Saaty's priority theory. Fuzzy Sets Syst. **11**, 199–227 (1983)
264. R. Csutora, J.J. Buckley, Fuzzy hierarchical analysis: the Lambda-Max method. Fuzzy Sets Syst. **120**, 181–195 (2001)
265. D. Taylor, S.A. Myers, A. Clauset, M.A. Porter, P.J. Mucha, Eigenvector-based centrality measures for temporal networks. Multiscale Model. Simul. **15**(1), 537–574 (2017)
266. D.G. Luenberger, Y. Ye, *Linear and Nonlinear Programming* (Springer Science & Business Media, 2008)
267. Y. Gong, N. Hu, J. Zhang, G. Liu, J. Deng, Multi-attribute group decision making method based on geometric Bonferroni mean operator of trapezoidal interval type-2 fuzzy numbers. Comput. Ind. Eng. **81**, 167–176 (2015)
268. R. Alyamani, S. Long, The application of fuzzy Analytic Hierarchy Process in sustainable project selection. Sustainability **12**(20), 8314 (2020)
269. G. Büyüközkan, F. Göçer, Y. Karabulut, A new group decision making approach with IF AHP and IF VIKOR for selecting hazardous waste carriers. Measurement, 134 (2019)
270. C.L. Hwang, K. Yoon, *Multiple Attribute Decision Making: Methods and Applications* (Springer-Verlag, New York, 1981)
271. S. Nădăban, S. Dzitac, I. Dzitac, Fuzzy TOPSIS: a general view. Procedia Comput. Sci. **91**, 823–831 (2016)
272. C. Gupta, A. Jain, N. Joshi, Fuzzy logic in natural language processing–a closer view. Procedia Comput. Sci. **132**, 1375–1384 (2018)
273. N.A. Omoregbe, I.O. Ndaman, S. Misra, O.O. Abayomi-Alli, R. Damaševičius, A. Dogra, Text messaging-based medical diagnosis using natural language processing and fuzzy logic. J. Healthc. Eng. **2020**, 1–14 (2020)
274. I. Truck, M.A. Abchir. Natural language processing and fuzzy tools for business processes in a geolocation context. Adv. Artif. Intell. (2017)

275. F. Herrera, E. Herrera-Viedma, and L. Mart´ınez, A fuzzy linguistic methodology to deal with unbalanced linguistic term sets. IEEE Trans. Fuzzy Syst. **16**(2), 354–370 (2008)
276. D. Gupta, K. Vani, C.K. Singh, Using Natural Language Processing techniques and fuzzy-semantic similarity for automatic external plagiarism detection, in *2014 International Conference on Advances in Computing, Communications and Informatics (ICACCI)* (IEEE, 2014), pp. 2694–2699
277. J.I.P.I.N.G. Sun, F. Karray, O.T.M.A.N. Basir, M.O.H.A.M.E.D. Kamel, Fuzzy logic-based natural language processing and its application to speech recognition, in *3rd WSES International Conference on Fuzzy Sets & Systems* (2002), pp. 11–15
278. A. Londhe, P.V.R.D.P. Rao, Dynamic classification of sentiments from restaurant reviews using novel fuzzy-encoded LSTM. Int. J. Recent. Innov. Trends Comput. Commun. **10**(9), 112–24 (2022). https://doi.org/10.17762/ijritcc.v10i9.5714
279. B. PhaniRaghavaa, S.A. Kumarb, An improved chatbot for predicting disease and medicines using natural language processing with fuzzy logic (2022)
280. A. Habous, E.H. Nfaoui, A fuzzy logic and ontology-based approach for improving the CV and job offer matching in recruitment process. Int. J. Metadata, Semant. Ontol. **15**(2), 104–120 (2021)
281. B. Alekhya, R. Sasikumar, An ensemble approach for healthcare application and diagnosis using natural language processing. Cogn. Neurodyn. **16**(5), 1203–1220 (2022)
282. T. Oka, P. Patankar, S. Rege, M. Dixit, Text summarization of news articles, in *ICT Systems and Sustainability: Proceedings of ICT4SD 2021*, vol. 1 (Springer Singapore, 2022), pp. 441–450
283. N. Premakumara, C. Shiranthika, C. Shyalika, S. Bandara, Optimized Text Summarization method based on fuzzy logic, in *2022 2nd International Conference on Image Processing and Robotics (ICIPRob)* (IEEE, 2022), pp. 1–6
284. A.Y. Zaremarjal, D. Yiltas-Kaplan, S. Lazemi, Emotion extraction from text using fuzzy-deep neural network, in *Intelligent and Fuzzy Techniques for Emerging Conditions and Digital Transformation: Proceedings of the INFUS 2021 Conference, held August 24–26, 2021*, vol. 2 (Springer International Publishing, 2022), pp. 329–338
285. K. Howells, A. Ertugan, Applying fuzzy logic for sentiment analysis of social media network data in marketing. Procedia Comput. Sci. **120**, 664–670 (2017)
286. S. Vashishtha, S. Susan, Fuzzy rule based unsupervised sentiment analysis from social media posts. Expert Syst. Appl. **138**, 112834 (2019)
287. H. Liu, M. Cocea, Fuzzy rule-based systems for interpretable sentiment analysis, in *2017 Ninth International Conference on Advanced Computational Intelligence (ICACI)* (IEEE, 2017), pp. 129–136
288. D. Jayashree, O. Pandithurai, S. Prasad, A.S. Suresh, S. Vigneshwaran, Sentimental analysis on voice based reviews using fuzzy logic, in *2021 International Conference on Advancements in Electrical, Electronics, Communication, Computing and Automation (ICAECA)* (IEEE, 2021), pp. 1–6
289. V. Ivanov, A review of fuzzy methods in automotive engineering applications. Eur. Transp. Res. Rev. **7**(3), 1–10 (2015)
290. P. Balachandra, Fault detection and recovery for automotive embedded system using rough set techniques, in *Information and Communication Technology for Sustainable Development: Proceedings of ICT4SD 2016*, vol. 1 (Springer, Singapore, 2018), pp. 233–241
291. K. Ouyang, Y. Li, B. Yang, Research on anti-collision control of pure electric vehicles, in *Proceedings of the 2019 International Conference on Robotics, Intelligent Control and Artificial Intelligence* (2019), pp. 637–642
292. A. Bonfitto, S. Feraco, M. Rossini, F. Carlomagno, Fuzzy logic method for the speed estimation in all-wheel drive electric racing vehicles. Commun. Sci. Lett. Univ. Zilina **23**(2), B117–B129 (2021)
293. J. Sun, X. Xue, K.W.E. Cheng, Fuzzy sliding mode wheel slip ratio control for smart vehicle anti-lock braking system. Energies **12**(13), 2501 (2019)
294. Z. Géczi, E. Tóth-Laufer, Fuzzy-based braking system model in driver assisted technology, in *2021 IEEE 15th International Symposium on Applied Computational Intelligence and Informatics (SACI)* (IEEE, 2021), pp. 000311–000316

295. X. Feng, J. Hu, Discrete fuzzy adaptive PID control algorithm for automotive anti-lock braking system. J. Ambient. Intell. Hum. Comput., 1–10 (2021)
296. K. Kim, Y. Kim, S. Kwak, Vision sensor based fuzzy system for intelligent vehicles. Sensors **19**(4), 855 (2019)
297. H.Y. Li, Velocity planning of automatic driving based on fuzzy neural network, in *2022 5th International Conference on Advanced Electronic Materials, Computers and Software Engineering (AEMCSE).* (IEEE, 2022), pp. 470–474
298. Songkroh A, Fooprateepsiri R, Lilakiataskun W, An intelligent risk detection from driving behavior based on BPNN and fuzzy logic combination, in *2014 IEEE/ACIS 13th International Conference on Computer and Information Science (ICIS)* (IEEE, 2014), pp. 105–110
299. S. Çeven, A. Albayrak, R. Bayır, Real-time range estimation in electric vehicles using fuzzy logic classifier. Comput. Electr. Eng. **83**, 106577 (2020)
300. D. Phan, A. Bab-Hadiashar, M. Fayyazi, R. Hoseinnezhad, R.N. Jazar, H. Khayyam, Interval type 2 fuzzy logic control for energy management of hybrid electric autonomous vehicles. IEEE Trans. Intell. Veh. **6**(2), 210–220 (2020)
301. B. Tifour, M. Boukhnifer, A. Hafaifa, C. Tanougast, Monitoring and control of energy management system for fuel cell hybrid in electrical vehicle using fuzzy approach. Diagnostyka **21** (2020)
302. M.A.A. Viegas, C.T. da Costa Jr, Fuzzy logic controllers for charging/discharging management of battery electric vehicles in a smart grid. J. Control., Autom. Electr. Syst. **32**(5), 1214–1227 (2021)
303. H. Mu, X. Li, J. Tang, Study with fuzzy self-turning PID controller on idle speed controlling of automobile, in *2017 Chinese Automation Congress (CAC)* (IEEE, 2017), pp. 5755–5759
304. Z. Al-Saadi, D. Phan Van, A. Moradi Amani, M. Fayyazi, S. Sadat Sajjadi, D. Ba Pham, ..., H. Khayyam, Intelligent driver assistance and energy management systems of hybrid electric autonomous vehicles. Sustainability **14**(15), 9378 (2022)
305. S. Arif, J. Iqbal, S. Munawar, Design of embedded motion control system based on modified fuzzy logic controller for intelligent cruise-controlled vehicles, in *2012 International Conference of Robotics and Artificial Intelligence* (IEEE, 2012), pp. 19–25
306. B. Zhang, H. Zhao, X. Zhang, Adaptive variable domain fuzzy PID control strategy based on road excitation for semi-active suspension using CDC shock absorber. J. Vib. Control., 10775463231152287 (2023)
307. Y.C. Lin, H.L.T. Nguyen, Adaptive neuro-fuzzy predictor-based control for cooperative adaptive cruise control system. IEEE Trans. Intell. Transp. Syst. **21**(3), 1054–1063 (2019)
308. F. Junyao, X. Wenping, L. Guohai, Vibration control for vehicle active suspension based on ANFIS method, in *2017 36th Chinese Control Conference (CCC)* (IEEE, 2017), pp. 9602–9606
309. M.M.M. Fahmy, A fuzzy algorithm for scheduling non-periodic jobs on soft real-time single processor system. Ain Shams Eng. J. **1**(1), 31–38 (2010)
310. M. Awadalla, A. Ahmad, S. Al-Busaidi, Scheduling of dependent real-time tasks using fuzzy logic technique. Am. Acad. Sch. Res. J. **6**(4) (2014)
311. S.R. Behera, N. Panigrahi, S.K. Bhoi, M. Bilal, K.S. Sahoo, D. Kwak, A distributed fuzzy optimal decision making strategy for task offloading in edge computing environment. IEEE Access **11**, 33189–33204 (2023)
312. K. Kolomvatsos, Data-driven type-2 fuzzy sets for tasks management at the edge. IEEE Trans. Emerg. Top. Comput. Intell. **6**(2), 377–386 (2021)
313. H.S. Behera, R. Pattanayak, P. Mallick, An improved fuzzy-based CPU scheduling (IFCS) algorithm for real time systems. Int. J. Soft Comput. Eng. (IJSCE) **2**(1), 2231–2307 (2012)
314. R.A. Kulkarni, S.H. Patil, N. Balaji, Efficient fuzzy based real time scheduling of workflows on cloud. Int. J. Eng. Adv. Technol. (IJEAT) **8**(6), 221–225 (2019)
315. H.E. Hassan, G. Nagib, K.H. Ibrahiem, A novel task scheduling approach for dependent non-preemptive tasks using fuzzy logic. IET Comput. Digital Tech. **15**(3), 214–222 (2021)
316. D.G. Harkut, M.S. Ali, Adaptive fuzzy hardware scheduler for real time operating system. Int. J. Comput. Digit. Syst. **5**(06) (2016)

317. M.T. Ogedengbe, M.A. Agana, New fuzzy techniques for real-time task scheduling on multiprocessor systems. Int. J. Comput. Trends Technol. **47**(3), 189–196 (2017)
318. K. Chrysafiadi, Improving task scheduling by using a fuzzy reasoner. Intell. Decis. Technol. **14**(2), 165–170 (2020)
319. K. Chrysafiadi, A fuzzy task scheduling method, in *Advances in Core Computer Science-Based Technologies: Papers in Honor of Professor Nikolaos Alexandris* (2021), pp. 305–323
320. Y. Hajoui, O. Bouattane, M. Youssfi, E. Illoussamen, New hybrid task scheduling algorithm with fuzzy logic controller in grid computing. Int. J. Adv. Comput. Sci. Appl. **9**(8) (2018)
321. C.J. Watkins, P. Dayan, Q-learning. Mach. Learn. **8**, 279–292 (1992)
322. M. Dorigo, M. Birattari, T. Stutzle, Ant colony optimization. IEEE Comput. Intell. Mag. **1**(4), 28–39 (2006)
323. H. Ben Alla, S. Ben Alla, A. Touhafi, A. Ezzati, A novel task scheduling approach based on dynamic queues and hybrid meta-heuristic algorithms for cloud computing environment. Clust. Comput. **21**(4), 1797–1820 (2018)
324. Y. Hou, L. Zhao, H. Lu, Fuzzy neural network optimization and network traffic forecasting based on improved differential evolution. Futur. Gener. Comput. Syst. **81**, 425–432 (2018)
325. T. Thamaraimanalan, C. Venkatesan, M. Ramkumar, A. Sivaramakrishnan, M. Marimuthu, ANFIS-based multilayered algorithm for botnet detection, in *2023 International Conference on Recent Advances in Electrical, Electronics, Ubiquitous Communication, and Computational Intelligence (RAEEUCCI)* (IEEE, 2023), pp. 1–5
326. P. David Collins, N. Chukwuchekwa, E. Longinus Sunday, Dynamic load balancing for congestion avoidance using adaptive neuro-fuzzy inference system in mobile communication network. Int. J. Electr. Comput. Eng. Syst. **13**(3), 183–193 (2022)
327. A.H. Hamamoto, L.F. Carvalho, L.D.H. Sampaio, T. Abrão, M.L. Proença Jr., Network anomaly detection system using genetic algorithm and fuzzy logic. Expert Syst. Appl. **92**, 390–402 (2018)
328. A. Chaudhary, M. Misra, A. Sardana, An efficient fuzzy controller based technique for network traffic classification to improve QoS, in *Proceedings of the Fifth International Conference on Security of Information and Networks* (2012), pp. 95–102
329. V.T. Nguyen, T.X. Nguyen, T.M. Hoang, N.L. Vu, A new anomaly traffic detection based on fuzzy logic approach in wireless sensor networks, in *Proceedings of the 10th International Symposium on Information and Communication Technology* (2019), pp. 205–209
330. A. Salama, R. Saatchi, D. Burke, Adaptive sampling technique using regression modelling and fuzzy inference system for network traffic (2017)
331. G. Raiesh, C. Swetha, R. Privanka, R. Vaishnavi, Congestion control in 6Lo WPAN networks using fuzzy logic (FLCC), in *2017 Ninth International Conference on Advanced Computing (ICoAC)* (IEEE, 2017), pp. 369–374
332. K. Theja, Intelligent traffic management in high-speed networks by fuzzy logic control. Int. J. Inf. Comput. Technol. **4**(12), 1175–1183 (2014)
333. M.F. Farooqui, M. Muqeem, A. Sultan, J. Nazeer, H.A. Abdeljaber, A Fuzzy Logic based Solution for Network Traffic Problems in Migrating Parallel Crawlers. Int. J. Adv. Comput. Sci. Appl. **14**(2) (2023)
334. J. Shreyas, H. Singh, S. Tiwari, N.N. Srinidhi, S.M. Dilip Kumar, CAFOR: congestion avoidance using fuzzy logic to find an optimal routing path in 6LoWPAN networks. J. Reliab. Intell. Environ., 1–16 (2021)
335. M. Vargheese, S. Vanithamani, D.S. David, G.R.K. Rao, Design of fuzzy logic control framework for QoS routing in manet. Intell. Autom. Soft Comput. **35**(3), 3479–3499 (2023)
336. A. Yazdinejad, A. Dehghantanha, R.M. Parizi, G. Srivastava, H. Karimipour, Secure intelligent fuzzy blockchain framework: Effective threat detection in IoT networks. Comput. Ind. **144**, 103801 (2023)
337. S.A. Wagan, J. Koo, I.F. Siddiqui, N.M.F. Qureshi, M. Attique, D.R. Shin, A fuzzy-based duo-secure multi-modal framework for IoMT anomaly detection. J. King Saud Univ. Comput. Inf. Sci. **35**(1), 131–144 (2023)

338. B.A. Bensaber, C.G.P. Diaz, Y. Lahrouni, Design and modeling an Adaptive Neuro-Fuzzy Inference System (ANFIS) for the prediction of a security index in VANET. J. Comput. Sci. **47**, 101234 (2020)
339. D.K. Jana, R. Ghosh, Novel interval type-2 fuzzy logic controller for improving risk assessment model of cyber security. J. Inf. Secur. Appl. **40**, 173–182 (2018)
340. J. Mo, B. Huang, X. Cheng, C. Huang, F. Wei, Improving security and stability of ad hoc on-demand distance vector with fuzzy neural network in vehicular ad hoc network. Int. J. Distrib. Sens. Netw. **14**(10), 1550147718806193 (2018)
341. S.A. Soleymani, A.H. Abdullah, M. Zareei, M.H. Anisi, C. Vargas-Rosales, M.K. Khan, S. Goudarzi, A secure trust model based on fuzzy logic in vehicular ad hoc networks with fog computing. IEEE Access **5**, 15619–15629 (2017)
342. İ Atacak, An ensemble approach based on fuzzy logic using machine learning classifiers for android malware detection. Appl. Sci. **13**(3), 1484 (2023)
343. H.I.H. Alsaadi, R.M. Almuttari, O.N. Ucan, O. Bayat, An adapting soft computing model for intrusion detection system. Comput. Intell. **38**(3), 855–875 (2022)
344. A. Chhabra, V. Vashishth, D.K. Sharma, A fuzzy logic and game theory based adaptive approach for securing opportunistic networks against black hole attacks. Int. J. Commun Syst **31**(4), e3487 (2018)
345. F.Q. Li, R.J. Zhao, S.L. Wang, L.B. Chen, A.W.C. Liew, W. Ding, Online intrusion detection for internet of things systems with full bayesian possibilistic clustering and ensembled fuzzy classifiers. IEEE Trans. Fuzzy Syst. **30**(11), 4605–4617 (2022)
346. M.B. Mansour, T. Abdelkader, M. Hashem, E.S.M. El-Horbaty, An integrated three-tier trust management framework in mobile edge computing using fuzzy logic. Peer J. Comput. Sci. **7**, e700 (2021)
347. C. Pham, L.A. Nguyen, N.H. Tran, E.N. Huh, C.S. Hong, Phishing-aware: a neuro-fuzzy approach for anti-phishing on fog networks. IEEE Trans. Netw. Serv. Manage. **15**(3), 1076–1089 (2018)
348. K.P. Vijayakumar, K. Pradeep Mohan Kumar, K. Kottilingam, T. Karthick, P. Vijayakumar, P. Ganeshkumar, An adaptive neuro-fuzzy logic based jamming detection system in WSN. Soft Comput. **23**, 2655–2667 (2019)
349. A. Janarthanan, D. Kumar, R.R. Antony, C.D. Parvathe, IADF security: insider attack detection using fuzzy logic in wireless multimedia sensor networks. Soft. Comput. **24**, 13893–13902 (2020)
350. G. Arulkumaran, R.K. Gnanamurthy, Fuzzy trust approach for detecting black hole attack in mobile adhoc network. Mob. Netw. Appl. **24**, 386–393 (2019)
351. A.R. Kumar, A. Sivagami, Fuzzy based malicious node detection and security-aware multipath routing for wireless multimedia sensor network. Multimed. Tools Appl. **79**, 14031–14051 (2020)
352. M. MaleknasabArdakani, M.A. Tabarzad, M.A. Shayegan, Detecting sybil attacks in vehicular ad hoc networks using fuzzy logic and arithmetic optimization algorithm J. Supercomput. **78**(14), 16303–16335 (2022)
353. R.K. Chahal, S. Singh, Fuzzy rule-based expert system for determining trustworthiness of cloud service providers. Int. J. Fuzzy Syst. **19**, 338–354 (2017)
354. W. Liu, Y. Mao, L. Ci, F. Zhang, A fuzzy approach to user-level intrusion detection. Int. J. Fuzzy Syst. **23**, 862–877 (2021)
355. J.B. Awotunde, F.E. Ayo, R. Panigrahi, A. Garg, A.K. Bhoi, P. Barsocchi, A multi-level random forest model-based intrusion detection using fuzzy inference system for internet of things networks. Int. J. Comput. Intell. Syst. **16**(1), 31 (2023)
356. M.I. Tariq, S. Ahmed, N.A. Memon, S. Tayyaba, M.W. Ashraf, M. Nazir, ..., M.M. Balas, Prioritization of information security controls through fuzzy AHP for cloud computing networks and wireless sensor networks. Sensors **20**(5), 1310 (2020)
357. G.P. Ramalingam, R.A. Xavier Annie, S. Gopalakrishnan, Optimized fuzzy enabled semi-supervised intrusion detection system for attack prediction. Intell. Autom. Soft Comput. **32**(3) (2022)

358. N. Naik, P. Jenkins, B. Kerby, J. Sloane, L. Yang, Fuzzy logic aided intelligent threat detection in cisco adaptive security appliance 5500 series firewalls, in *2018 IEEE International Conference on Fuzzy Systems (FUZZ-IEEE)* (IEEE, 2018, July), pp. 1–8
359. A. Alabdulatif, I. Khalil, H. Kumarage, A.Y. Zomaya, X. Yi, Privacy-preserving anomaly detection in the cloud for quality assured decision-making in smart cities. J. Parallel Distrib. Comput. **127**, 209–223 (2019)
360. B. Sujata, P. Ravi Kiran Varma, Combining fuzzy C-means and KNN algorithms in performance improvement of intrusion detection system, in *Proceedings of International Conference on Computational Intelligence and Data Engineering: ICCIDE 2017* (Springer, Singapore, 2018), pp. 359–370
361. Z. Rustam, A.S. Talita, Fuzzy kernel robust clustering for anomaly based intrusion detection, in *2018 Third International Conference on Informatics and Computing (ICIC)* (IEEE, 2018, October), pp. 1–4
362. L. Decker, D. Leite, L. Giommi, D. Bonacorsi, Real-time anomaly detection in data centers for log-based predictive maintenance using an evolving fuzzy-rule-based approach, in *2020 IEEE International Conference on Fuzzy Systems (FUZZ-IEEE)* (IEEE, 2020, July), pp. 1–8
363. https://www.gps.gov/systems/gnss/. Accessed 12 Apr 2023
364. https://www.geotab.com/blog/what-is-gps/. Accessed 12 Apr 2023
365. http://www.aerostudents.com/courses/avionics/InertialNavigationSystems.pdf. Accessed 12 Apr 2023
366. D. Titterton, J.L. Weston, J. Weston, *Strapdown Inertial Navigation Technology*, vol. 17 (IET, 2004)
367. H. Nourmohammadi, J. Keighobadi, Fuzzy adaptive integration scheme for low-cost SINS/GPS navigation system. Mech. Syst. Signal Process. **99**, 434–449 (2018)
368. X. Wang, X. Wang, J. Zhu, F. Li, Q. Li, H. Che, A hybrid fuzzy method for performance evaluation of fusion algorithms for integrated navigation system. Aerosp. Sci. Technol. **69**, 226–235 (2017)
369. S. Khankalantary, S. Rafatnia, H. Mohammadkhani, An adaptive constrained type-2 fuzzy Hammerstein neural network data fusion scheme for low-cost SINS/GNSS navigation system. Appl. Soft Comput. **86**, 105917 (2020)
370. E.S. Abdolkarimi, G. Abaei, A. Selamat, M.R. Mosavi, A hybrid type-2 fuzzy logic system and extreme learning machine for low-cost INS/GPS in high-speed vehicular navigation system. Appl. Soft Comput. **94**, 106447 (2020)
371. C. Wang, L. Wang, J. Qin, Z. Wu, L. Duan, M. Cao, …, Q. Wang, Development of a vision navigation system with fuzzy control algorithm for automated guided vehicle, in *2015 IEEE International Conference on Information and Automation* (IEEE, 2015), pp. 2077–2082
372. K. Teo, B. Goh, O.K. Chai, Fuzzy docking guidance using augmented navigation system on an AUV. IEEE J. Oceanic Eng. **40**(2), 349–361 (2014)
373. S. Pan, X. Xu, L. Zhang, Y. Yao, A novel SINS/USBL tightly integrated navigation strategy based on improved ANFIS. IEEE Sens. J. **22**(10), 9763–9777 (2022)
374. O. Petrova, G. Tabunshchyk, T. Kapliienko, O. Kapliienko, Fuzzy verification method for indoor-navigation systems, in *2018 14th International Conference on Advanced Trends in Radioelecrtronics, Telecommunications and Computer Engineering (TCSET)* (IEEE, 2018), pp. 65–68
375. Z. Yue, B. Lian, Y. Gao, Robust adaptive filter using fuzzy logic for tightly-coupled visual inertial odometry navigation system. IET Radar Sonar Navig. **14**(3), 364–371 (2020)
376. G.G. Kalach, G.P. Kalach, Navigation system based on the fuzzy logic expert system. Int. J. Adv. Trends Comput. Sci. Eng. **8**(6), 2693–2698 (2019)
377. S. Rafatnia, H. Nourmohammadi, J. Keighobadi, Fuzzy-adaptive constrained data fusion algorithm for indirect centralized integrated SINS/GNSS navigation system. GPS Solut. **23**(3), 62 (2019)
378. E.S. Abdolkarimi, M.R. Mosavi, Wavelet-adaptive neural subtractive clustering fuzzy inference system to enhance low-cost and high-speed INS/GPS navigation system. GPS Solut. **24**(2), 36 (2020)

379. R. Bautista-Montesano, R. Bustamante-Bello, R.A. Ramirez-Mendoza, Explainable navigation system using fuzzy reinforcement learning. Int. J. Interact. Des. Manuf. (IJIDeM) **14**(4), 1411–1428 (2020)
380. C.H. Tseng, S.F. Lin, D.J. Jwo, Fuzzy adaptive cubature Kalman filter for integrated navigation systems. Sensors **16**(8), 1167 (2016)
381. M.A. Ali, M. Mailah, W.A. Jabbar, K. Moiduddin, W. Ameen, H. Alkhalefah, Autonomous road roundabout detection and navigation system for smart vehicles and cities using laser simulator–fuzzy logic algorithms and sensor fusion. Sensors **20**(13), 3694 (2020)
382. Y. H. Lin, C. M. Yu, C. Y. Wu, Towards the design and implementation of an image-based navigation system of an autonomous underwater vehicle combining a color recognition technique and a fuzzy logic controller. Sensors **21**(12), 4053 (2021)
383. X. Ding, H. Bian, H. Ma, R. Wang, Ship trajectory generator under the interference of wind current and waves. Sensors **22**(23), 9395 (2022)
384. T.K. Sethuramalingam, Fuzzy logic controller based ship navigation system. J. Phys.: Conf. Ser. **1850**(1), 012113 (IOP Publishing, 2021)
385. I.N. Stăncel, I. Petrescu, Fuzzy logic algorithm in autonomous vehicle navigation based on RFID sensors, in *2019 IEEE 25th International Symposium for Design and Technology in Electronic Packaging (SIITME)* (IEEE, 2019), pp. 70–73
386. C.H. Chen, C.J. Lin, S.Y. Jeng, H.Y. Lin, C.Y. Yu, Using ultrasonic sensors and a knowledge-based neural fuzzy controller for mobile robot navigation control. Electronics **10**(4), 466 (2021)
387. M. Lazreg, N. Benamrane, Hybrid system for optimizing the robot mobile navigation using ANFIS and PSO. Robot. Auton. Syst. **153**, 104114 (2022)
388. M.H. Haider, Z. Wang, A.A. Khan, H. Ali, H. Zheng, S. Usman, …, P. Zhi, Robust mobile robot navigation in cluttered environments based on hybrid adaptive neuro-fuzzy inference and sensor fusion. J. King Saud Univ.-Comput. Inf. Sci. **34**(10), 9060–9070 (2022)
389. A. Keshtkar, S. Arzanpour, An adaptive fuzzy logic system for residential energy management in smart grid environments. Appl. Energy **186**, 68–81 (2017)
390. E. Işık, M. Inallı, Artificial neural networks and adaptive neuro-fuzzy inference systems approaches to forecast the meteorological data for HVAC: the case of cities for Turkey. Energy **154**, 7–16 (2018)
391. C. Turhan, S. Simani, G.G. Akkurt, Development of a personalized thermal comfort driven controller for HVAC systems. Energy **237**, 121568 (2021)
392. P. Danassis, K. Siozios, C. Korkas, D. Soudris, E. Kosmatopoulos, A low-complexity control mechanism targeting smart thermostats. Energy Build. **139**, 340–350 (2017)
393. A.C. Duman, H.S. Erden, Ö. Gönül, Ö. Güler, A home energy management system with an integrated smart thermostat for demand response in smart grids. Sustain. Cities Soc. **65**, 102639 (2021)
394. H. Chaouch, C. Çeken, S. Arı, Energy management of HVAC systems in smart buildings by using fuzzy logic and M2M communication. J. Build. Eng. **44**, 102606 (2021)
395. I. Ganchev, A. Taneva, K. Kutryanski, M. Petrov, Decoupling fuzzy-neural temperature and humidity control in HVAC systems. IFAC-PapersOnLine **52**(25), 299–304 (2019)
396. V. Martínez Viol, E.M. Urbano González, M. Delgado Prieto, J.L. Romeral Martínez, HVAC early fault detection using a fuzzy logic based approach, in *Renewable Energy and Power Quality Journal (RE&PQJ)*, vol. 18. AEDERMACP (European Association for the Development of Renewable Energies and Power Quality) (2020), pp. 1–6
397. A. Ghulghule, H. Khandekar, Smart HVAC system using fuzzy logic, in *Smart Technologies for Energy, Environment and Sustainable Development: Select Proceedings of ICSTEESD 2018* (Springer, Singapore, 2019), pp. 41–51
398. A. Berouine, E. Akssas, Y. Naitmalek, F. Lachhab, M. Bakhouya, R. Ouladsine, M. Essaaidi, A fuzzy logic-based approach for HVAC systems control, in *2019 6th International Conference on Control, Decision and Information Technologies (CoDIT)* (IEEE, 2019), pp. 1510–1515
399. L. Hang, D.H. Kim, Enhanced model-based predictive control system based on fuzzy logic for maintaining thermal comfort in IoT smart space. Appl. Sci. **8**(7), 1031 (2018)

400. H. Meryem, B. Omar, R. Abdelhadi, K. Tajeddine, Development of an adaptive HVAC fuzzy logic controller for commercial facilities: a case study. Adv. Sci., Technol. Eng. Syst. J. **5**(4), 331–337 (2020)
401. C. Anastasiadi, A.I. Dounis, Co-simulation of fuzzy control in buildings and the HVAC system using BCVTB. Adv. Build. Energy Res. **12**(2), 195–216 (2018)
402. S. Banjar, A.I. Hussein, M. Mohamad, R.H.M. Aly, Design of fuzzy controller for a hybrid active/passive cooling system in smart homes with a windcatcher, in *2023 7th International Conference on Green Energy and Applications (ICGEA)* (IEEE, 2023), pp. 192–198
403. B. Omarov, A. Altayeva, Z. Suleimenov, Y. Im Cho, B. Omarov, Design of fuzzy logic based controller for energy efficient operation in smart buildings, in 2017 *First IEEE International Conference on Robotic Computing (IRC)* (IEEE, 2017), pp. 346–351
404. A. Behravan, B. Kiamanesh, R. Obermaisser, Fault diagnosis of DCV and heating systems based on causal relation in fuzzy Bayesian belief networks using relation direction probabilities. Energies **14**(20), 6607 (2021)
405. K. Parvin, M.A. Hannan, A.Q. Al-Shetwi, P.J. Ker, M.F. Roslan, T.I. Mahlia, Fuzzy based particle swarm optimization for modeling home appliances towards energy saving and cost reduction under demand response consideration. IEEE Access **8**, 210784–210799 (2020)
406. Y.E. Isikdemir, G. Erturk, H. Ates, M.O. Tas, Fuzzy inference and machine learning based HVAC control system for smart buildings, in *2022 Global Energy Conference (GEC)* (IEEE, 2022), pp. 116–119
407. S. Iqbal, M. Sarfraz, A.S. Allahloh, A. Nabi, Intelligent residential energy management via fuzzy logic and smart load simulator, in *2023 International Conference on Power, Instrumentation, Energy and Control (PIECON)* (IEEE, 2023), pp. 1–6
408. F. Klashanov, A. Swan, Application of fuzzy logic methods for controlling buildings HVAC equipment, in *IOP Conference Series: Materials Science and Engineering*, vol. 869, no. 2 (IOP Publishing, 2020), p. 022010
409. K.M. Reena, A.T. Mathew, L. Jacob, Real-time occupancy based HVAC control using interval type-2 fuzzy logic system in intelligent buildings, in *2017 12th IEEE Conference on Industrial Electronics and Applications (ICIEA)* (IEEE, 2017), pp. 455–460
410. K. Rajeswari Subramaniam, C.T. Cheng, T.Y. Pang, Fuzzy logic controlled simulation in regulating thermal comfort and indoor air quality using a vehicle heating, ventilation, and air-conditioning system. Sensors **23**(3), 1395 (2023)
411. S. Rahman, M. Rabbi, O. Altwijri, M. Alqahtani, T. Sikandar, I.I. Abdelaziz, ..., K. Sundaraj Fuzzy logic-based improved ventilation system for the pharmaceutical industry Int. J. Eng. Technol. **7**(2), 640–645 (2018)

Chapter 4
Comparative Discussion

Abstract Fuzzy logic is an artificial intelligence technique that allows reasoning in cases of uncertain, vague, and subjective data. It is a powerful tool, which provides a flexible, intuitive, and human-like reasoning way. It has application in various fields, like control systems, expert systems, decision-making systems, medical systems, navigation systems, image processing, security systems etc. In this chapter, a comparative discussion concerning the application of fuzzy logic in software systems in several fields is presented. Also, we examine and compare several fuzzy logic-based techniques, which are preferred to be used in fuzzy logic-based software systems.

Fuzzy logic is an artificial intelligence technique that allows reasoning in cases of uncertain, vague, and subjective data. It enables to calculate with words, describing data and phenomena with linguistic words and concerning several degrees of truth that are represented with values from 0 to 1. Therefore, fuzzy logic is a powerful tool, which provides a flexible, intuitive, and human-like reasoning way to model uncertain and imprecise data. It has application in various fields, like control systems, expert systems, decision-making systems, medical systems, navigation systems, image processing, security systems etc. In this chapter, we will compare:

- The total number of published scientific works, which concern fuzzy logic-based systems, until August 4, 2023, per field of application.
- The fuzzy logic-based techniques that are used in systems per field of application.

4.1 Fields of Application of Fuzzy Logic

In this section, the several fields of application of fuzzy logic for the development of expert and human-like reasoning systems are presented. The fields of application are:

- Medical systems
- Tutoring systems
- Decision making support and Recommended systems

K. Chrysafiadi, *Fuzzy Logic-Based Software Systems*, Learning and Analytics in Intelligent Systems 34, https://doi.org/10.1007/978-3-031-44457-9_4

- Natural language processing systems
- Automotive Systems
- Scheduling, Network Traffic and Security software systems
- Navigation systems
- Heating, Ventilation and Air Conditioning (HVAC) systems.

More details about the use of fuzzy logic in systems of these fields were given in the previous chapter.

4.2 Published Scientific Works Concerning the Use of Fuzzy Logic in Expert Systems in Several Fields

In this section, the scope is to answer to the following questions:

1. In which field of application is the fuzzy logic used the most?
2. When was the first scientific work published, concerning the use of fuzzy logic in systems in each field of application?
3. How did the use of fuzzy logic in systems progress the last five years in each field of application?
4. How much of the published works concern the year 2003 (until August 4, 2023) for each field of application?

To answer these questions, a search in the academic tool 'Scopus' was performed. The criteria for the search were:

- "fuzzy logic" and "medical"
- "fuzzy logic" and "tutoring system"
- "fuzzy logic" and ("decision making" or "recommended system")
- "fuzzy logic" and (natural language processing)
- "fuzzy logic" and "automotive systems"
- "fuzzy logic" and ("scheduling" or "network traffic" or "security")
- "fuzzy logic" and "navigation system"
- "fuzzy logic" and "HVAC".

For answering the above questions corresponding filters were applied to the results. The search results are summarized in the Table 4.1. Also, in Fig. 4.1, the published works concerning the use of fuzzy logic in total in each field of application are presented in a graphical way. Similarly, in Fig. 4.2 the published works concerning the use of fuzzy logic in each field of application from 2019 to 2023 are presented in a graphical way. Finally in Fig. 4.3 the published works concerning the use of fuzzy logic in each field of application in 2023 (until August 4, 2023) are presented in a graphical way.

From the above results, the following conclusions are derived:

- There is a significantly increased use of fuzzy logic in decision-making support systems and recommended systems.

Table 4.1 Summarized search results in Scopus concerning the use of fuzzy logic in several fields

Field	Year of first publication	Published works in total	Published works in the last 5 years (2019–2023)	Published works in 2023 (until August 4, 2023)	Percentage of published work in the last 5 years (2019–2023) (%)
Medical systems	1974	3,343	993	135	29.70
Tutoring systems	1993	98	25	5	25.51
Decision making support and recommended systems	1974	9,826	2,983	388	30.36
Natural language processing systems	1978	406	121	17	29.80
Automotive systems	1989	29	4	0	13.79
Scheduling, network traffic and security software systems	1979	4,838	1,633	197	33.75
Navigation systems	1988	555	90	5	16.22
heating, ventilation and air conditioning (HVAC) systems	1990	196	56	7	28.57

- Second in interest of use of fuzzy logic is the fields of scheduling, network traffic management and security software systems.
- There is, also, a satisfactory scientific interest in incorporating fuzzy logic in expert medical diagnostic systems and tutoring systems.
- The use of fuzzy logic in automotive systems does not receive much acceptance and research interest.
- The use of fuzzy logic in tutoring systems begins more recently in relation to other fields of application.
- The fields in which the use of fuzzy logic receive an increased interest in the last five years are: (i) scheduling, network traffic management and security software

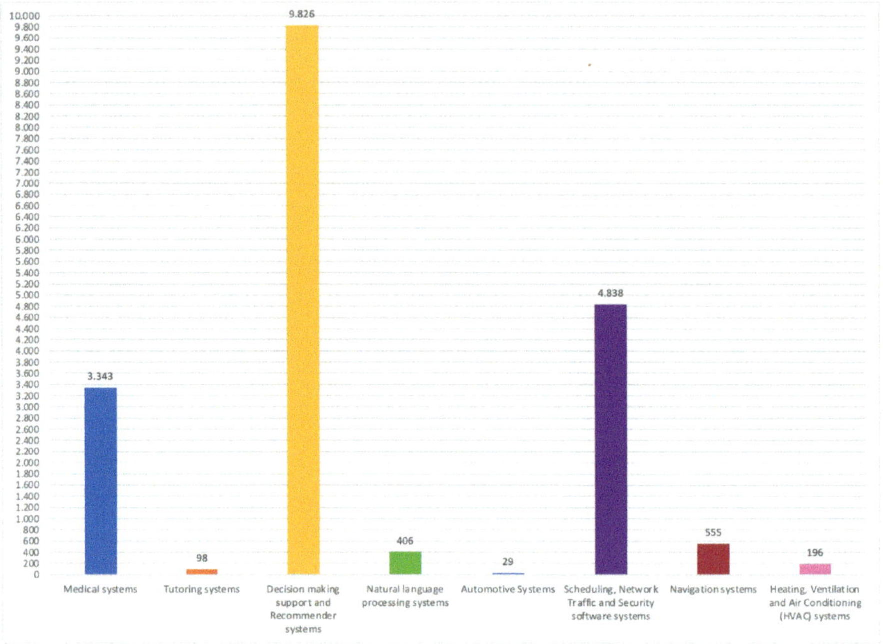

Fig. 4.1 Published studies concerning the use of fuzzy logic in systems of several fields in total

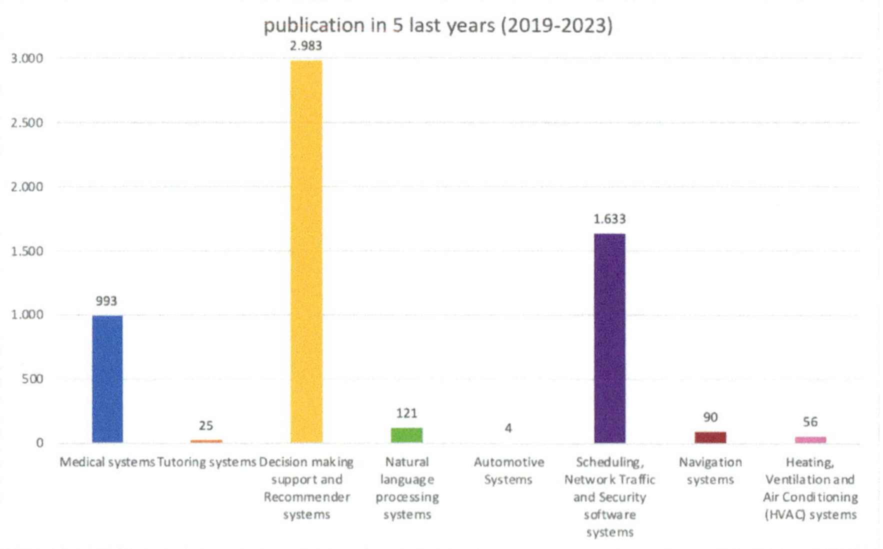

Fig. 4.2 Published studies concerning the use of fuzzy logic in systems of several fields in the last five years (2019–2023)

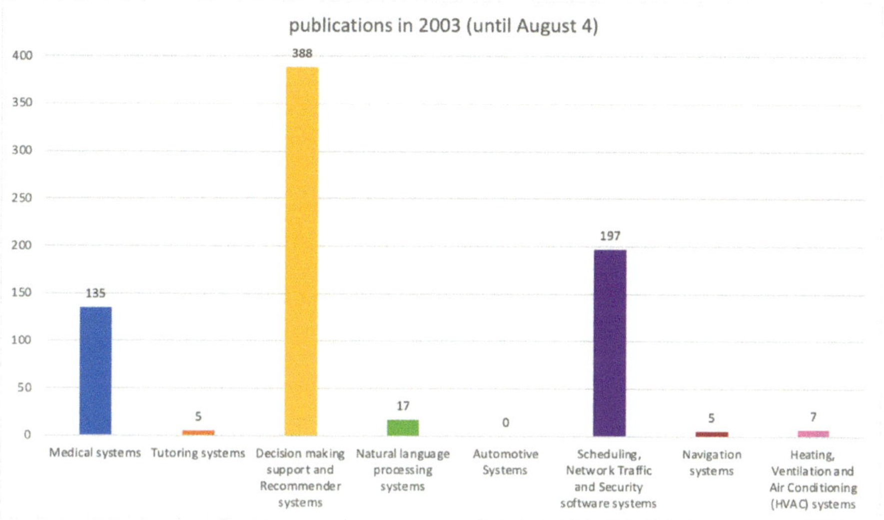

Fig. 4.3 Published studies concerning the use of fuzzy logic in systems of several fields in 2023 (until August 4, 2023)

systems, (ii) decision making support and recommended systems, (iii) natural language processing systems, (iv) medical systems, and (v) HVAC.

4.3 Fuzzy Logic Techniques and Methods that are Used in Systems in Several Fields

In this section, the fuzzy techniques that are used in systems in the different field of applications are presented (Table 4.2). For the needs of this comparative work, the following categories of fuzzy techniques and methods have been defined:

- Fuzzy clustering (fuzzy K-nearest neighbors clustering, Fuzzy C-means clustering, fuzzy support vector machine algorithm, fuzzy kernel robust clustering etc.)
- Fuzzy Inference System (Mamdani FIS, Takagi-Sugeno FIS)
- Fuzzy neural network
- Adaptive neuro fuzzy inference system
- Fuzzy decision tree
- Fuzzy cognitive map
- Fuzzy sets and linguistic modeling (intuitionistic fuzzy sets, interval type-2 fuzzy logic, fuzzy TOPSIS, interval-valued q-rung orthopair fuzzy sets, fuzzy AHP, hesitant fuzzy set, Q-rung orthopair fuzzy sets, Pythagorean fuzzy sets, etc.).

We notice the following:

Table 4.2 The applied fuzzy techniques and methods in systems of several fields

Fuzzy technique / Field	Fuzzy clustering	Fuzzy inference system	Fuzzy neural network	ANFIS	Fuzzy decision tree	Fuzzy cognitive map	Fuzzy sets and linguistic modeling
Medical systems	•	•	•	•	•	•	•
Tutoring systems	•	•	•	•	•	•	•
Decision making support and recommended systems	•	•	•	•	•	•	•
Natural language processing systems	•	•	•				•
Automotive systems		•	•	•			•
Scheduling, network traffic and security software systems	•	•	•	•	•		•
Navigation systems	•	•	•	•	•	•	•
Heating, ventilation and air conditioning (HVAC) systems	•	•	•	•		•	•

- There are not scientific works concerning the use of ANFIS, fuzzy decision trees and fuzzy cognitive maps in natural processing language systems.
- There are not scientific works concerning the use of fuzzy clustering methods, fuzzy decision trees and fuzzy cognitive maps in automotive systems.
- There is not scientific work concerning the use of fuzzy cognitive maps in scheduling, network traffic and security software systems.
- There is not scientific work concerning the use of fuzzy decision trees in HVAC systems.
- In all other fields of application all the aforementioned categories of fuzzy techniques have been used.

Furthermore, in Table 4.3 the percentages of the total published studies that concern systems that integrate each category of fuzzy techniques for each field of application are presented. Also, in Table 4.4 the percentages of the published studies, from 2019 until 2023 (until 4, August 2023), that concern systems that integrate each category of fuzzy techniques for each field of application are presented. These percentages were calculated regarding the results of searches with corresponding keywords and filters in the academic tool 'Scopus'. According to them, the following conclusions are derived:

- The most common-used technique in fuzzy logic-based medical systems are fuzzy clustering, fuzzy inference system, and fuzzy neural networks.
- The most common-used technique in fuzzy logic-based tutoring systems, decision making support, fuzzy logic-based recommended systems, and fuzzy logic-based automotive systems are fuzzy sets, rules and linguistic modeling.
- In the last five years there is an increased interest in using fuzzy inference systems in fuzzy logic-based tutoring systems.
- In the last five years there is not any published work that concerns the use of fuzzy clustering, fuzzy inference system, fuzzy neural network, ANFIS, fuzzy decision tree, and fuzzy cognitive map in automotive systems.

Table 4.3 The percentages of the total published studies that concern systems that integrate each category of fuzzy techniques for each field of application

Fuzzy technique Field	Fuzzy clustering (%)	Fuzzy inference system (%)	Fuzzy neural network (%)	ANFIS (%)	Fuzzy decision tree (%)	Fuzzy cognitive map (%)	Fuzzy sets and linguistic modeling (%)
Medical systems	29.76	20.07	20.67	11.49	1.74	5.33	10.94
Tutoring systems	8.16	8.16	12.25	2.04	1.02	6.12	62.25
Decision making support and recommended systems	6.5	12.29	10.12	4.87	1.51	6.85	57.86
Natural language processing systems	10.10	8.87	15.02	3.70	1.72	3.94	43.35
Automotive systems	0	6.90	10.35	3.45	0	0	79.3
Scheduling, network traffic and security software systems	14.28	19	17.51	9.38	1.01	2.94	35.88
Navigation systems	3.24	29.37	14.60	15.50	0.72	1.44	35.13
Heating, ventilation and air conditioning (HVAC) systems	2.55	14.29	17.86	10.71	0	2.04	52.55

Table 4.4 The percentages of the published studies, from 2019 until 2023 (until 4, August 2023), that concern systems that integrate each category of fuzzy techniques for each field of application

Fuzzy technique / Field	Fuzzy clustering (%)	Fuzzy inference system (%)	Fuzzy neural network (%)	ANFIS (%)	Fuzzy decision tree (%)	Fuzzy cognitive map (%)	Fuzzy sets and linguistic modeling (%)
Medical systems	23.25	21.37	26.76	13.26	1.64	3.74	9.98
Tutoring systems	0	20	12	0	0	4	64
Decision making support and Recommended systems	6.37	17	14.11	6.71	0.87	8.18	53.24
Natural language processing systems	13.22	16.53	27.27	12.40	1.65	4.13	24.8
Automotive systems	0	0	0	0	0	0	100
Scheduling, network traffic and security software systems	15.13	23.15	26.52	12.80	1.25	3.12	18.03
Navigation systems	0	28.03	24.24	16.67	3.08	0.76	27.22
Heating, ventilation and air conditioning (HVAC) systems	1.79	25	28.57	19.64	0	1.79	23.21

- The most common-used technique in fuzzy logic-based natural language processing systems are fuzzy sets, rules and linguistic modeling concerning the published studies of all years.
- In the last five years, however, there is an increased interest in using fuzzy neural networks, which is the prevalent fuzzy logic-based technique, in fuzzy logic-based natural language processing systems.
- The most common-used technique in fuzzy logic-based scheduling, network traffic and security software systems are fuzzy sets, rules and linguistic modeling concerning the published studies of all years.
- In the last five years, however, there is an increased interest in using fuzzy neural networks, and fuzzy inference systems, which are the prevalent fuzzy logic-based techniques, in fuzzy logic-based scheduling, network traffic and security software systems.
- The most common-used technique in fuzzy logic-based navigation systems are fuzzy sets, rules and linguistic modelling, and fuzzy inference system.

- In the last five years, there is an increased interest in using fuzzy neural networks in fuzzy logic-based navigation systems.
- The most common-used technique in HVAC systems are fuzzy sets, rules and linguistic modeling concerning the published studies of all years.
- In the last five years, however, there is an increased interest in using fuzzy neural networks, and fuzzy inference systems, which are the prevalent fuzzy logic-based techniques, in fuzzy logic-based HVAC systems.
- Concerning all the years, the most common-used fuzzy technique in fuzzy logic-based software systems is fuzzy sets, rules and linguistic modeling.
- Concerning the last five years (2019–2023), the most common-used fuzzy techniques in fuzzy logic-based software systems are fuzzy neural networks and fuzzy inference systems.

Chapter 5
Conclusions and Discussion

Abstract Although, fuzzy logic, which was introduces in 1960s by Dr. Lotfi Zadeh, was not well received by the scientific community, nowadays it is used extensively in smart applications and systems. Its ability to complete computations with linguistic elements (such as simple words) and handle uncertainty, imprecision and subjectivity makes fuzzy logic a valuable tool. It is used in many automated systems and processes in several disciplines like medicine, education, decision making and recommendation, natural language processing, automotive engineering and industry, heating, ventilation and air-conditioning, navigation, scheduling, network traffic and security. In this chapter, the role of fuzzy logic in smart systems and applications and its significant contribution to their efficiency, reliability and robustness are discussed. Except of the advantages of fuzzy logic, its limitations are, also, presented. Furthermore, the fuzzy logic-based techniques, which are used most commonly in software systems are presented. In addition, how artificial intelligence benefits from fuzzy logic is discussed. Finally, the areas in which there is a great scope of research and further development, concerning the use of fuzzy logic-based models and techniques in smart systems, are presented.

Although, fuzzy logic, which was introduces in 1960s by Dr. Lotfi Zadeh, was not well received by the scientific community, nowadays it is used extensively in smart applications and systems. Its ability to complete computations with linguistic elements (such as simple words) and handle uncertainty, imprecision and subjectivity makes fuzzy logic a valuable tool. It is used in medical systems for disease diagnosis, image analysis and processing; in tutoring systems for handling the uncertain and imprecise student characteristics and data providing personalized tutoring and educational services; in decision making support and recommended systems for comparing alternatives and suggesting the optimal solution; in natural language processing for improving recognition of ambiguous words; in automotive systems for assisted driving and improved efficient management of energy consumption; in task scheduling; in managing network traffic; in improving the performance of security software systems enhancing the process of threats and anomalies detection; in navigation systems improving the tracking process and navigation accuracy; in HVAC systems allowing the adaptive adjustment of indoor climate parameters.

K. Chrysafiadi, *Fuzzy Logic-Based Software Systems*, Learning and Analytics in Intelligent Systems 34, https://doi.org/10.1007/978-3-031-44457-9_5

The widespread use of fuzzy logic in smart applications and systems is, mainly, due to its potent ability to model and reason in situations where data is uncertain, incomplete and subjective. Many applications and systems, which have been built to facilitate our lives by performing automatically several processes, deal with real world scenarios problems, which usually confront with uncertainty, vagueness and data that cannot be handed by binary logic effectively. Therefore, fuzzy logic is the ideal solution for make these applications and systems accurate, effective and robust. Fuzzy logic enables systems for making decisions based on subjective data and several degrees of truth making them to think and reason like humans. Furthermore, the ability of fuzzy logic to handle data with linguistic terms, enables the systems to recognize and model the human language and knowledge in a better way. This results to a more effective human–machine interactions in applications where take as input user queries, like home assistants or natural language processing systems. Regarding the above, AI gains from the fuzzy logic, which offers great benefits to smart systems.

Although, the first application of fuzzy logic in smart systems concerned control systems and making-decision support systems in industry, the use of fuzzy logic received significant recognition in many other areas, like medical systems, scheduling, network traffic and security software systems. In fact, in the last five years (2019–2023), the use of fuzzy logic has increased remarkably in the areas of scheduling, network traffic and security software systems, natural language processing systems medical systems, HVAC systems and tutoring systems. More research is needed on the use of fuzzy logic in automotive and navigation systems.

Regarding the techniques of fuzzy logic, which are used more frequently and commonly, they are: fuzzy mathematical models, several types of fuzzy sets and fuzzy rules, fuzzy clustering methods, several types of fuzzy inference systems, fuzzy neural networks, ANFIS, fuzzy logic-based decision trees, and fuzzy cognitive maps. All these fuzzy logic-based techniques have been used in systems of all the aforementioned fields of applications. Only concerning natural language processing systems and automotive systems there is a restricted number of fuzzy logic-based techniques that have been used in them. Therefore, there is a great scope for research in the use of different fuzzy logic-based techniques, like ANFIS and fuzzy cognitive maps, in natural language processing systems and automotive systems. Furthermore, the most extensive-used fuzzy logic-based techniques are fuzzy mathematical models, like interval type-2 fuzzy logic; several types of fuzzy sets and fuzzy rules, like intuitionistic fuzzy sets, hesitant fuzzy sets and Pythagorean fuzzy sets; and linguistic modelling methods, like fuzzy TOPSIS and fuzzy AHP. However, in the last five years (2019–2023), fuzzy neural networks and fuzzy inference systems have been added to the most extensive-used fuzzy logic method in smart applications and systems.

Consequently, fuzzy logic began as a mathematical framework, but it has been developed in an advanced technology that contributes to the efficiency and reliability of smart applications and systems in several areas. Its widespread usage is real-world-proof for the benefits that fuzzy logic offers to systems. Advantages of fuzzy logic include: (i) interpretability, (ii) flexibility, (iii) human-like reasoning, (iv) control in nonlinear systems, (v) managing of multiple conflicting or overlapping data or information, and (vi) simple integration of experts' knowledge. However, fuzzy logic has, also, some limitations that include: (i) lack of standardization; (ii) requiring to acquire, interpret and integrate experts' knowledge, and (iii) subjectivity in defining fuzzy sets, membership functions and fuzzy rules. Nevertheless, the major ability of fuzzy logic to handle incomplete and uncertain data constitutes a powerful tool for many applications and systems. Also, the combination of fuzzy logic with other AI techniques, like machine learning and artificial neural networks, contributes to significant mitigation of the above limitations of fuzzy logic. To conclude, fuzzy logic is a valuable tool, which is used extensively in software applications and systems to make them more intelligent, efficient, reliable, and robust. Its further development and the research on the use of fuzzy logic-based models and techniques in smart systems have great potentials in the future.

Appendix A
Examples of Fuzzy Logic–Based Systems in Several Areas

K. Chrysafiadi, *Fuzzy Logic-Based Software Systems*, Learning and Analytics in Intelligent Systems 34, https://doi.org/10.1007/978-3-031-44457-9

Table A.1 Examples of diagnotic fuzzy logic−based medical systems

References	Disease	Fuzzy AI algorithm	FIS	Input	Input membership functions	Output	Output membership functions	Defuzzification method
[6]	Lupus nephritis	Fuzzy K-nearest neighbors	–	Population size, search space di-mensions, the maximum number of function evaluations, objective function, upper and lower bounds of search space	–	The optimal value and optimal solution of the objective function	–	–
[7]	Cerebral palsy	–	Mamdani	8 parameters that were associated with kinematic data	Gaussian	Gait pattern	Triangular	Middle of maxima
[8]	Breast cancer	Fuzzy C−means	Takagi–Sugeno	6 features of the density maps of mammograms	Bell−shaped	Density class	Not defined	The maximum value
[9]	Cardiac tachyarrhythmias	–	Not specified	RR−interval, onset and instability	Triangular	6 diagnostic categories: ventricular fibrillation, ventricular tachycardia, sinus tachycardia, detection artifacts and irregularities (including extrasystoles), atrial tachyarrhythmias and no tachycardia (i.e. normal sinus rhythm)	Singletons	–
[10]	Cardiac arrhythmia	Fuzzy KNN algorithm	Mamdani	Integration matrix of heartbeat classifications	Trapezoidal	Heartbeat classification	Trapezoidal	Centroid

(continued)

Table A.1 (continued)

References	Disease	Fuzzy AI algorithm	FIS	Input	Input membership functions	Output	Output membership functions	Defuzzification method
[11]	Chronic kidney disease	–	Mamdani	16 diagnostic parameters: age, systolic blood pressure, diastolic blood pressure, proteinuria, albuminuria, creatinine, Glomerular filtration rate, blood urea nitrogen, parathyroid hormone, hemoglobin, calcium, phosphorus, polar height to pole in ultrasound, history of hypertension, history of diabetes, and family history of CKD	Gaussian, trapezoidal, triangular, and singleton	1 variable CKD (chronic kidney disease)	Not specified	Not specified
[12]	Knee osteoarthritis	–	Mamdani	Feature scoring for each method and category	Triangular and trapezoidal	Fuzzy feature importance	Triangular and trapezoidal	Not specified
[13]	Parkinson	Fuzzy k-nearest neighbor	–	Feature set	–	Possibility of parkinson	–	–

(continued)

Table A.1 (continued)

References	Disease	Fuzzy AI algorithm	FIS	Input	Input membership functions	Output	Output membership functions	Defuzzification method
[14]	Diabetes mellitus	Fuzzy support vector machine	–	8 diagnostic parameters: number of pregnancies, plasma glucose concentration, blood pressure, skin fold thickness, insulin, body mass index, diabetes pedigree function, age	–	Significant features: plasma glucose concentration, Body Mass Index and fuzzy rules	–	–
[15]	Coronary heart disease	–	Mamdani	7 parameters: cholesterol, blood pressure, physical activity, smoking, age, BMI, and diabetes	Trapezoidal and triangular	Presence of heart disease (disease_classification)	Trapezoidal and triangular	Not specified
[14]	Rheumatic–musculoskeletal disease	Fuzzy concept map	Mamdani	28 diagnostic parameters	Triangular	Diagnosis of rheumatic–musculoskeletal disease	Triangular	Centre of gravity
[17]	Cervical cancer	–	Mamdani	Age, onset of sexual activity, sexual couples, number of pregnancies, first pregnancy age, sexually transmitted diseases, cervical lesions	Trapezoidal and triangular	Risk	Triangular	Centroid
		–	Mamdani	Nucleus dimension, nuclei staining, homogeneous nucleation, nucleolus presence, risk	Trapezoidal and triangular	Injury resolution	Trapezoidal	Centroid

(continued)

Table A.1 (continued)

References	Disease	Fuzzy AI algorithm	FIS	Input	Input membership functions	Output	Output membership functions	Defuzzification method
[18]	Chronic kidney disease	–	Mamdani	Nephron functionality, blood sugar, diastolic blood pressure, systolic blood pressure, age, body mass index, and smoke	Trapezoidal	Patient's condition	Triangular and trapezoidal	Centroid
[19]	Hypertension	Neuro fuzzy system	Mamdani	Age, body mass index, blood pressure, heart rate	Triangular and trapezoidal	Risk of hypertension	Triangular	Centroid
[20]	Renal cancer	ANFIS	–	Smoking, dialysis, occupational exposure, genetic or hereditary	Gaussian	Existence of cancer stage	Gaussian	Not specified
		–	Mamdani	Blood in urine, red blood cell count, flank pain, tumor size, Von Hippel–Lindau gene, high blood pressure, and trichloroethylene exposure		Cancer stage		
[21]	Cardiovascular disease	–	Hierarchical fuzzy inference system	Layer 1: blood oxygen, color lips	Triangular	Layer 1: y1	Triangular	Not specified
				Layer 2: breathing rate, y1		Layer 2: y2		
				Layer 3: heart rate, y2		Layer 3: risk level		
[22]	Cardiovascular disease	–	Mamdani	Systolic blood pressure, body mass index, triglycerides	Triangular	risk	Triangular	Not specified

(continued)

Table A.1 (continued)

References	Disease	Fuzzy AI algorithm	FIS	Input	Input membership functions	Output	Output membership functions	Defuzzification method
[23]	Lung cancer	–	Takagi–sugeno	Anamnesis, doubling time, the degree of the smokers, age and performance	Triangular and trapezoidal	Cancer existence	Not specified	Weighted average
[24]	Knee diseases	–	Mamdani	Data for knee inflammatory diseases: pain, age, swelling, deformity, restriction, ESR, bilateral swelling, bilateral tenderness,, medial joint line tenderness	Triangular and trapezoidal	Osteoarthritis, rheumatoid arthritis, osteonecrosis	Triangular and trapezoidal	Centroid
[25]	Respiratory diseases	–	Mamdani	Fever, headache, myalgia, fatigue, nasal congestion, sneezing, sore throat, difficulty breathing, rhinorrhoea	Sigmoidal and generalized bell	COVID−19 diagnosis, allergies diagnosis, common cold diagnosis and influenza diagnosis	Not specified	Not specified
[26]	Heart disease	–	Mamdani	8 attributes of a patient	Triangular and trapezoidal	Diagnosis of heart disease	Triangular and trapezoidal	Not specified
[27]	Cardiovascular diseases	–	Mamdani	Heart rate, patient age, systolic blood pressure, diastolic blood pressure, sex, chest pain, diabetes, smoking, electrocardiogram and total cholesterol	Triangular	Risk of having a cardiovascular disease	Triangular	Centroid method

(continued)

Table A.1 (continued)

References	Disease	Fuzzy AI algorithm	FIS	Input	Input membership functions	Output	Output membership functions	Defuzzification method
[28]	Osteoporosis	–	Mamdani	Sex, age, body mass index, amount of paint, heredity, calcium deficiency, physical stress, years since menopause and consumption alcohol or cigarette	Trapezoidal and triangular	Osteoporosis prediction	Trapezoidal	Not specified
[29]	Anemia	–	Mamdani	Content of serum iron in the blood, total iron binding capacity of blood serum, serum ferritin content, reticulocyte count	Triangular	Type of anemia	Triangular	Not specified
[30]	Coronary heart disease	–	Mamdani	Age, total cholesterol, LDL cholesterol, HDL cholesterol, systolic blood pressure, diastolic blood pressure, sex, smoking, diabetes	Trapezoidal and triangular	Coronary heart disease risk level	Triangular	Centre of gravity
[31]	Covid–19	–	Mamdani	White blood cell, lymphocyte, monocytes, locate dehydrogenase	Trapezoidal and triangular	Covid–19 diagnosis value	Trapezoidal and triangular	Centre of gravity

* The reference numbers concern the references in Chap. 3

Table A.2 Examples of fuzzy logic–based image analysis and/or segmentation medical systems

References	Disease	Aim	Fuzzy AI algorithm	Input	Output
[32]	Breast cancer	Detection	Fuzzy C–means	Contrast, energy, homogeneity, entropy and correlation	Boundary by system
[33]	Breast cancer	Contrast enhancement of mammograms	A fuzzy–based technique that consists of fuzzification, modification of the membership function, and defuzzification	A mammogram image	Enhanced image
[34]	Several diseases	Enhance diagnosis result by medical images	Interval–valued intuitionistic fuzzy set	Medical image	Fused image
[35]	Breast cancer	Medical image segmentation for cancer detection	Intuitionistic possibilistic fuzzy c–mean	Mammogram images	Image objects
[36]	Brain tumor	MRI segmentation	Interval type–II fuzzy logic	Optimal cluster boundaries	Tutor and non-tutor regions
[38]	Lung cancer	Image segmentation	Fuzzy C–means	Lung image	Classification of the lung cancer images as normal image and disease affected image
[40]	Breast cancer	Ultrasound image segmentation	Fuzzy clustering and fuzzy level set algorithm	Ultrasound image	Segmented ultrasound breast cancer image
[41]	Lung cancer	Prediction	Fuzzy neural network	Features from the enhanced segmented CT image	Lung cancer diagnosis
[42]	Lung diseases	Medical image segmentation	Fuzzy C–means	Medical image	Segmented image
[43]	Cancer	Segmentation	Fuzzy C–means	Microscopic biopsy images	Segmented microscopic biopsy images
[44]	Glioma brain	Brain images classification	ANFIS	Extracted from fused brain image features	Glioma and non–glioma images
[46]	Glioma brain	Glioma brain tumor segmentation	Fuzzy C–means	Magnetic resonance images (MRI)	Segmented MRI
[50]	Lung nodule	Computed tomography images	Fuzzy C–means	Computed tomography images (CT)	Segmented CT

(continued)

Table A.2 (continued)

References	Disease	Aim	Fuzzy AI algorithm	Input	Output
[51]	Abdominal aortic aneurysm	Segmentation	Fuzzy C–means	Magnetic resonance image (MR)	Binarized image
[52]	Cataract	Edge estimation	Fuzzy inference system	Human eye medical image	The location of the edges
[53]	Brain tumor	Classification	ANFIS	Extracted features from medical image	Abnormal and normal brain tumor
[54]	Brain lesions	Segmentation	Fuzzy C–means	Brain MR images	Common knowledge among different clustering tasks
[55]	Covid–19	Detection	ANFIS	Data of a X–ray image: contrast, correlation, energy, homogeneity, mean, variance	Existence of disease
[56]	Covid–19	Diagnosis	Fuzzy neural network	Various image formats of lungs	Disease classification

* The reference numbers concern the references in Chap. 3

Table A.3 Examples of fuzzy logic–based student models

References	Aim	Student's characteristics to model	Fuzzy technique	Reason for fuzzy logic usage
[63]	Provide the appropriate learning material and exercises	Knowledge level	Fuzzy rules	Define the complexity of the next quiz
[64]	Offer personalized learning path	Knowledge level, misconceptions, and cognitive transitions	Mamdani FIS	Identify the student's cognitive state and adapt the learning path to her/his learning abilities and pace
[65]	Provide an effective learning path	Knowledge level and learning behavior	Fuzzy descriptors and rules	Computational analysis of the student's characteristics and learning behaviors
[66]	Improve the prediction accuracy in student model design	Students' performance	Six trapezoidal fuzzy sets and their membership functions	To handle uncertainty
[67]	Offering adaptive education in foreign languages	Students' knowledge and performance	Mamdani FIS	Provide a personalized study plan
[68]	The determination of the learner's knowledge level in French according to her/his knowledge level in English, and vice versa	Knowledge level and knowledge dependencies among concepts of English and French	Fuzzy rules	The adaptation of the instructional content
[69]	Determine the student's learning level	Student's MYCIN certain Factor and the number of attempts to solve a problem	Mamdani FIS	Provide personalized learning material
[70]	The dynamic delivery of the learning objects tailored to learners' needs	Student's knowledge level	Fuzzy rules	Determine the number of exercises concerning each teaching grammatical issue that have to be delivered to the learner
[71]	Adjudge best tutoring strategy for the learner	Student's knowledge level and learning style	Mamdani FIS	Decide about the learning strategy that suits better to the learner's needs
[72]	Provide an adaptation of the course material	Student's performance in test	Mamdani FIS	Define the degree of student's success in a particular domain topic
[73]	To identify Learner characteristics and preferences of the learning–disabled learners	Student's score and time responding on a pre–test	Fuzzy neural network	Define the severity of student's learning difficulties and their strength
[74]	Provide the most appropriate learning environment	Student's attentiveness, excitement and test completion	Fuzzy rules	To validate the selected learning environment

(continued)

Table A.3 (continued)

References	Aim	Student's characteristics to model	Fuzzy technique	Reason for fuzzy logic usage
[75]	Generate math exercises that meet the student's needs	Student's performance on solving exercises (error, help, time) and her/his emotion	Mamdani FIS	To define the difficulty level of the next exercise
[76]	To predict student's learning style	The eight dimensions of the four learning styles of the Felder Silverman	Fuzzy decision tree	To define the student's learning style
[77]	To adapt the type of learning material	The student's preferences for a certain resource category	Fuzzy Neural Network	To detect the student's learning style

* The reference numbers concern the references in Chap. 3

Table A.4 Examples of fuzzy logic–based student's performance prediction systems

References	Aim	Fuzzy technique	Input	Output
[78]	To facilitate educational institutions to identify deficiencies and improve learning outcomes and educational quality	Takagi–Sugeno FIS	Students' scores and the number of absences	The predicted grade
[79]	Student classification	Fuzzy neural network with Takagi–sugeno FIS	Interest, talent and motivation	Classification model
[80]	To predict student performance	Fuzzy cognitive map	Student performance antecedents	Student overall performance
[81]	To predict student performance	Fuzzy neural network	Students' score in the first and the second semester	Student's predicted performance
[82]	To predict student performance	Mamdani FIS	Daily engagement in studies, participation, english literacy, previous educational achievement, prior knowledge assessment and related course score	Student course performance
[83]	To predict student performance	Fuzzy neural network with multiple parameters set	Academic and student data	The student's future performance

* The reference numbers concern the references in Chap. 3

Table A.5 Examples of fuzzy logic—based student's evaluation performance systems

References	Aim	Fuzzy technique	Input	Output
[84]	Evaluating students' performance	Mamdani FIS	Class record and exam performance	Student performance
[85]	Evaluating students' learning performance	Mamdani FIS	Importance, complexity, and difficulty of each test item	The student's comprehensive level
[86]	Evaluating students' performance	Mamdani FIS	Lecture evaluation, practical evaluation, self—work of the student with the teacher, final evaluation	The overall student's evaluation
[87]	Student performance evaluation	Mamdani FIS	Students' marks in three tests	The final student's performance
[88]	Evaluation of students' intelligence level	Fuzzy C—means and Takagi–Sugeno FIS	The student's grades in three semesters	The student's academic performance
[89]	Student academic performance analysis	Fuzzy C—means	Students' grades	Four clusters related to students' performance
[90]	Dividing the population of students into homogeneous groups according to their performance	Fuzzy C—means and Takagi–Sugeno FIS	Students' grades	Clusters related to students' performance and the overall student's performance
[91]	To evaluate the learners' performance of practice courses	Fuzzy cognitive map	Survey results and exam grades	The critical tasks and concepts measuring the success of the practical courses
[92]	Ranking students' performance	Fuzzy sets and fuzzy numbers	Linguistic evaluation values on 13 criteria and the importance of them	The distance of student's evaluation to the highest rank
[93]	To evaluate students' project according to several criteria	Mamdani FIS	24 criteria concerning knowledge level, technical and presentation knowledge and skills	The student's score on 5 main assessment criteria and the final overall score

* The reference numbers concern the references in Chap. 3

Table A.6 Examples of fuzzy logic–based systems for adaptive assessment

References	Aim	Fuzzy technique	Reason for fuzzy logic usage
[98]	Create adaptive assessment tests	Membership functions for linguistic data representation	Model domain expert's knowledge
[99]	Test complexity testing	Fuzzy rules	Define the complexity level of the next block of questions
[100]	Create dynamically adaptive tests	Fuzzy representation of knowledge level and fuzzy rules	Recognize the forgotten, partially known, overall known and unknown concepts
[101]	Create adaptive assessment test	Fuzzy rules	Define the number and the difficulty level of the test items for each level of the Revised Bloom Taxonomy
[103]	Selection of the next test item	Mamdani FIS	Estimate the student's ability
[104]	Select the most appropriate test items	Mamdani FIS	Calculate the activity's relevance degree

* The reference numbers concern the references in Chap. 3

Table A.7 Examples of fuzzy logic−based systems for decision making

References	Field	Aim	Fuzzy technique
[219]	Education	Higher business education evaluation	A Choquet integral−based hesitant fuzzy GLDS (HF−GLDS) method
[222]	Education	Modeling systems thinking in action among higher education leaders	Fuzzy set theory
[224]	Education	To choose the lecture that offers the highest achievement in an adaptive web−based educational system	Rule−based fuzzy cognitive maps
[220]	Education	Prioritization of students' requirements in higher education	Fuzzy TOPSIS
[221]	Education	Program evaluation and management policy	Fuzzy extent analysis method
[217]	Education	Online education platform performance evaluation	Interval−valued q−rung orthopair fuzzy sets
[218]	Education	Online education live platform selection	Linguistic intuitionistic cubic fuzzy set
[216]	Education	Selection of teaching and research staff in higher education	Fuzzy TOPSIS
[200]	ERP system selection	Rank a set of critical success factors related to ERP implementation process	Fuzzy AHP
[198]	ERP system selection	Evaluating ERP system	q−rung orthopair fuzzy sets
[201]	ERP system selection	Selection among ERP outsourcing alternatives	Fuzzy AHP
[171]	Industry	Selection of the best technology forecasting in aerospace industry	Fuzzy MUTLIMOORA
[167]	Industry	Evaluating manufacturing plant locations	Fuzzy inference system
[166]	Industry	Selection of lean tools	Fuzzy QFD
[163]	Industry	Warehouse site selection	Fermetean fuzzy sets
[172]	Manufacturing	To assess and prioritise the enablers of manufacturing flexibility	Fuzzy AHP and fuzzy TOPSIS
[174]	Industry	Valuation of enterprise management innovation	Fuzzy AHP
[165]	Industry	Performance measurement model for Industry 4.0 in small and medium−sized manufacturing firms	Fuzzy AHP
[164]	Industry	Selecting a solar panel company	Fuzzy TOPSIS
[170]	Manufacturing	Minimizing completion time in manufacturing process in supply chains	Fuzzy TOPSIS
[168]	Industry	Evaluation of green lean production	Fuzzy COPRAS method
[162]	Manufacturing	Manufacturing resources allocation	Intuitionistic fuzzy sets

(continued)

Table A.7 (continued)

References	Field	Aim	Fuzzy technique
[173]	Industry	Evaluation and selection of shrimp suppliers in the fisheries industry	Fuzzy analytical network process
[211]	Investment	Implementation and effective resources integration of potential regional energy internet	Interval type–2 fuzzy number based Choquet integral
[209]	Investment	Examining the influence of contextual factors on investment decision–making	Fuzzy AHP
[213]	Investment	Investment decision making	Probabilistic hesitant fuzzy set
[210]	Investment	To find an appropriate system for microgeneration energy investments and identify optimal renewable energy alternatives	Q–rung orthopair fuzzy sets
[207]	Investment	Investment decision making	Interval valued intuitionistic fuzzy soft set
[206]	Investment	Investment decision making	Fuzzy inference system
[205]	Investment	Investment selection	Intuitionistic fuzzy multiple attribute decision-making model
[120]	Logistics	To identify the order of establishing temporary logistics hubs	Fuzzy linguistic variables and fuzzy set theory
[122]	Logistics	Prioritizing reverse logistics adoption barriers	Fuzzy AHP
[121]	Logistics	Select a separate location for collection and distribution	Fuzzy DEMATEL and fuzzy ANP and fuzzy AHP
[123]	Logistics	Logistics outsourcing	Multihesitant fuzzy linguistic term set
[118]	Logistics	Third–party logistics selection in sustainable supply chain	Fuzzy AHP and fuzzy VIKOR
[119]	Logistics	Third–party logistics provider selection	Fuzzy AHP
[117]	Logistics	Selection of a third–party reverse logistics service provider	Intuitionistic fuzzy MCDM methods
[183]	Medical	To prioritize the level of significance of the elements causing challenges in the Internet of medical things	Fuzzy AHP and fuzzy TOPSIS
[176, 182, 187]	Medical	Medical diagnosis	Fuzzy soft sets
[184]	Medical	Medical diagnosis	Pythagorean fuzzy soft matrix
[181]	Medical	Medical diagnosis	Arithmetic of generalised parabolic fuzzy numbers
[188]	Medical	Medical diagnosis	Arithmetic triangular C
[186]	Medical	Medical diagnosis	Intuitionistic fuzzy sets

(continued)

Table A.7 (continued)

References	Field	Aim	Fuzzy technique
[191]	Risk management	Evaluation of the critical success factors of dynamic enterprise risk management	q−ROF−entropy−RS−ARAS" approach (based on the q−rung orthopair fuzzy set)
[194]	Risk management	Selecting irrigation modernization strategies	Fuzzy simple additive weighting (FSAW) approach
[192]	Risk management	Risk management of photovoltaic power plants	Fuzzy theory sets
[196]	Risk management	Evaluating financial risk management	Fuzzy TOPSIS
[193]	Risk management	Cloud risk management	Linguistic fuzzy sets and interval numbers
[195]	Risk management	Analysis of flood risk management	Interval−valued intuitionistic fuzzy numbers
[142]	Supplier selection	Sustainable supplier selection	A Choquet integral based interval Type−2 trapezoidal fuzzy multiple attribute method
[153]	Supplier selection	Sustainable supplier selection	Interval−valued Pythagorean fuzzy set
[157]	Supplier selection	Service supplier selection	Fuzzy AHP
[144]	Supplier selection	Selection of reliable suppliers for a company	HFLTS
[150]	Supplier selection	Green supplier selection	Fuzzy AHP and fuzzy TOPSIS and fuzzy ELECTRE
[141]	Supplier selection	Supplier selection for construction project	Fuzzy hypersoft sets
[143]	Supplier selection	Wind turbine supplier selection in a renewable energy project	spherical fuzzy sets
[155]	Supplier selection	Green suppliers' selection	Interval−valued Pythagorean fuzzy numbers
[151]	Supplier selection	Raw materials supplier selection in plastics industry	Fuzzy analytic network process
[146]	Supplier selection	Green supplier selection in the coffee bean supply chain	Fuzzy AHP
[145]	Supplier selection	Supplier selection in healthcare supply chain management during the COVID−19 pandemic	Fuzzy rough numbers
[124]	Sustainability	Selection of sustainability issues that are most relevant for creating shared value for both business and society	Fuzzy AHP
[131]	Sustainability	Assessment of sustainability indicators for green building manufacturing	Fuzzy DEMATEL
[134]	Sustainability	Sustainability assessment of wastewater treatment technologies	Fuzzy TOPSIS and FSWARA andFMOORA
[135]	Sustainability	Sustainability assessment of alternative marine fuels	Fuzzy TOPSIS and fuzzy logarithmic least squares
[236]	Sustainability	Ranking tourist destinations and evaluating their performances under a sustainability perspective	Fuzzy AHP

(continued)

Table A.7 (continued)

References	Field	Aim	Fuzzy technique
[237]	Sustainability	Prioritizing the flare technologies in the oil, gas, and chemical plants	Fuzzy SWARA and fuzzy WASPAS
[233]	Sustainability	Green supplier selection	Fuzzy AHP and Fuzzy TOPSIS
[232]	Sustainability	Healthcare sustainability evaluation	Fuzzy set theory
[236]	Technology	Optimal selection of clean technologies	Fuzzy AHP
[158]	Technology	Adoption of smart manufacturing technologies by small and medium–sized enterprises	Fuzzy TOPSIS
[239]	Technology	Optimal selection of Iron and Steel wastewater treatment technology	Fuzzy AHP and Fuzzy TOPSIS
[228]	Technology	Selection of digital technologies	Fuzzy AHP
[235]	Technology	Evaluation of existing data collection technologies	Fuzzy AHP
[226]	Technology	Evaluation and selection of information technologies for order picking in a warehouse system	IMF SWARA–Z and fuzzy EDAS–Z
[230]	Technology	Select the appropriate software for cloud computing system implementation	Interval–Valued Intuitionistic Fuzzy sets
[238]	Technology	Performance evaluation in value engineering	Fuzzy AHP
[231]	Technology	Renewal of technology for healthcare equipment	Fuzzy AHP
[232]	Technology	Selection of emerging technologies	Pythagorean fuzzy numbers
[227]	Technology	Selection of appropriate communication technology for smart grid	Fuzzy TOPSIS
[111]	Transportation	Prioritization of unmanned aerial vehicles in transportation system	Fuzzy Level Based Weight Assessment and Fuzzy Rough Hamacher Combined Compromise Solution
[110]	Transportation	Transportation strategy	Interval–valued complex fuzzy sets
[109]	Transportation	Route selection in multimodal transportation networks	Fuzzy AHP
[108]	Transportation	To find most preferable transportation mode	Interval type–2 fuzzy variables
[106]	Transportation	Pavement strategy selection	Hesitant Fuzzy methods

* The reference numbers concern the references in Chap. 3

Table A.8 Examples of fuzzy logic–based natural language processing systems

References	Field	Aim	NPL system	Fuzzy technique
[273]	Medical	To understand user's symptoms, which are entered via text messages and diagnose the disease	Chatbot	Mamdani FIS with weighted fuzzy rules
[274]	Geolocation	To interpret some vague questions concerning geolocation into business processes	Customer service application	2–tuple fuzzy linguistic representation model
[276]	Detection of plagiarism	To compare academia and scientific research documents and detect, automatically, plagiarism	Semantic Search application	Fuzzy–semantic similarity measure method
[277]	Speech recognition	To predict the words failed to be recognized by a speech recognition system	Speech recognition system	Fuzzy linguistic modelling of the words and fuzzy rules
[278]	Sentiment analysis	To identify customers' feelings through reviews	Sentiment analysis system	Fuzzy sets with triangular membership functions and fuzzy rules
[279]	Medical	To predict disease and medicine	Chatbot	Fuzzy support vector machine algorithm
[280]	Text matching	To match a CV with description of offered job positions	Semantic Search application	Mamdani FIS
[281]	Medical	To integrate patients' healthcare records from heterogeneous sources and offer diagnose services	A medical knowledge extraction web–based application	Mamdani FIS
[282]	Text summarization	To generate summary of documents	Semantic analysis system	Mamdani FIS
[283]	Text summarization	To generate summary of documents	Semantic analysis system	Mamdani FIS
[284]	Emotion recognition	To extracts emotions from textual data in social media	Semantic analysis system	Fuzzy neural network
[285]	Sentiment analysis	To analyze the consumers' perceptions from text and emoji comments in social media	Social bot	Mamdani FIS
[286]	Sentiment analysis	To define the users' sentiment from social media posts	Bot applications	Mamdani FIS
[287]	Sentiment analysis	To recognize users' emotions and opinions	–	Sugeno–Tsukamoto FIS
[288]	Sentiment analysis		Sentiment analysis application	Fuzzy sets and linguistic representation

* The reference numbers concern the references in Chap. 3

Table A.9 Examples of fuzzy logic–based automotive systems

References	Field	Aim	Fuzzy technique	Combined technique
[290]	Assisted driving	Fault detection and recovery	Fuzzy rough sets	Interface extended finite–state machine approach
[291]	Assisted driving	Anti–collision control	Mamdani FIS	–
[292]	Assisted driving	To estimate the vehicle's speed in all wheels	Mamdani FIS	–
[293]	Assisted driving	To yield anti–lock control of wheels with an adaptive sliding surface	Takagi–Sugeno FIS	–
[294]	Assisted driving	Intelligent control of the braking system	Takagi–Sugeno FIS	–
[295]	Assisted driving	Control of anti-lock braking system	Fuzzy rule–based system	–
[296]	Assisted driving	Estimation of the risk level of collision with a pedestrian	Mamdani FIS	Computer vision technologies
[297]	Assisted driving	To estimate the vehicle's acceleration when it faces obstacles	Fuzzy neural network	–
[298]	Driver modelling	Detection and identification of the driving behavior	Mamdani FIS	Back propagation neural network
[299]	Energy efficiency	Range estimation in electric vehicles	Mamdani FIS	–
[300]	Energy efficiency	Energy consumption management	Interval type 2 fuzzy logic system	–
[301]	Energy efficiency	Energy management for hybrid electrical vehicles	Takagi–Sugeno FIS	Particle swarm optimization
[302]	Energy efficiency	Energy management of electric vehicles	Mamdani FIS	–
[303]	Energy efficiency	To control vehicle's idle speed	Linguistic values and fuzzy rules	–
[305]	Fuel–energy efficiency	Vehicle's speed controlling	Mamdani FIS	–
[306]	Ride comfort	Improvement of the vibration reduction effect	Mamdani FIS	–
[308]	Ride comfort	Controlling the vibration of a vehicle suspension system	ANFIS	Particle swarm optimization

(continued)

Table A.9 (continued)

References	Field	Aim	Fuzzy technique	Combined technique
[307]	Ride comfort and fuel efficiency	Maintenance of the desired distance between two vehicles and efficient fuel consumption	Takagi–Sugeno FIS and adaptive neuro–fuzzy predictor–based control	–
[304]	Ride comfort and fuel efficiency	Controlling acceleration and manage fuel consumption	ANFIS and Tankagi–Sugeno model	Model predictive control

* The reference numbers concern the references in Chap. 3

Table A.10 Examples of fuzzy logic–based task scheduling

Refs	System	Aim	Fuzzy technique	Input	Output
[309]	Real–time single processor system	To minimize average delay, number of overdue jobs, and throughput times of jobs	Mamdani FIS	1st FIS: the task exterior priority, the task waiting time and the task processing time 2nd FIS: the task processing priority and the task deadline time	1st FIS: the task processing priority 2nd FIS: the task final priority
[310]	Battery–powered real time systems	To reduce energy consumption	Mamdani FIS	(i) The worst case execution time, (ii) deadline, (iii) blocking time, (iv) arriving time	(i) task priority and (ii) processor speed
[311]	Edge computing environment	To make decision about task offloading	Mamdani FIS	(i) Task size, (ii) network bandwidth, (iii) network delay, (iv) energy consumption, and (v) mobile edge computing virtual machine utilization	Offloading decision
[312]	Edge computing environment	To eliminate the latency in the delivery of responses and to save the bandwidth of the network	Type–2 fuzzy logic system	(i) The room for execution of additional tasks in a node, (ii) the speed that the task is delivered at pre–defined intervals, (iii) the load that the task will cause to the processing node	Potential of allocation
[313]	Real time systems	The efficient usage of CPU and minimizing tasks' throughput, waiting time and turnaround time	Mamdani FIS	(i) The task's burst time, (ii) response ration, (iii) the task's priority	New calculated priority of the task
[314]	Cloud computing environment	To manage uncertainty of virtualization and migration and to succeed better management of tasks' deadlines	Mamdani FIS	(i) The expected execution time, (ii) the assigned priority, and (iii) the task deadline	The overall task priority
[315]	Multiprocessor systems	To succeed high performance	Mamdani FIS	(i) The height of the node, which represents the task, in the Directed Acyclic Graph, (ii) the task's execution time	The task's priority

(continued)

Table A.10 (continued)

Refs	System	Aim	Fuzzy technique	Input	Output
[316]	Multiprocessor real–time systems	To improve efficiency of the system	Mamdani FIS	1st FIS: the job exterior priority, the job processing priority, the job waiting time 2nd FIS: the job processing priority, the job worst case execution time	1st FIS: the job processing priority 2nd FIS: the job final priority
[317]	Multiprocessor real–time systems	To decrease the minimum response time, the turnaround time and the number of deadlines missed	Mamdani FIS	(i) The task's arrival time, (ii) computation time, (iii) deadline	The task's runtime priority for execution
[318]	Real–time systems	To minimize the mean waiting time and the mean turnaround time	Mamdani FIS	(i) Task's deadline, (ii) remaining execution time	The task's priority
[319]	Real–time systems	To enhance the system's performance	Mamdani FIS	(i) The task's execution time, (ii) waiting time	The task's priority
[320]	Distributed systems	To increase the efficiency in resource allocation	Mamdani FIS	(i) The node rapidity, (ii) the node state	The node reward
[323]	Cloud computing environment	To optimize cloud utilization and quality of service	Mamdani FIS	1st FIS: iteration parameter 2nd FIS: the task length, the CPU speed, the RAM memory and the status of the virtual machine	1st FIS: the inertia weight 2nd FIS: the fitness value

* The reference numbers concern the references in Chap. 3

Table A.11 Examples of fuzzy logic use in network traffic management

Refs	Aim	Fuzzy technique	Combine technique	Input	Output
[324]	To forecast network traffic	Fuzzy neural network	Differential evolution algorithm and BP algorithm	–	
[325]	To detect botnet	ANFIS	–	Network traffic data	Botnet detection
[326]	ANFIS to avoid congestion in mobile communication networks	ANFIS	–	Received signal strength, received signal quality, network traffic	Hysteresis value
[327]	To detect network anomalies and generate alarm	Gaussian fuzzy membership functions	Genetic algorithm	–	
[328]	Calculation of bandwidth consumption	Mamdani FIS	–	(i) The average of mean time between two consecutive packets arriving at the interface, (ii) the number of packets of a specific protocol detected, and (iii) the ratio of number of protocol packets to the total number of packets arriving	The percentage bandwidth consumption
[329]	To detect anomaly network traffic in wireless sensor networks	Mamdani FIS	–	Mean and variance of sensory data	Type of network traffic
[330]	Adaptive sampling	Mamdani FIS	Regression model	(i) The current duration value of inter–sampling interval and (ii) the Euclidean distance	The updated duration of inter–sample interval
[331]	Congestion controlling in 6LoWPAN networks	Mamdani FIS	–	(i) Buffer occupancy, (ii) queue length, (iii) packet priority, (iv) packet size,	transmission rate
[332]	intelligent network traffic management	Mamdani FIS	–	The queue error deviation and the integration of queue error	The aggregate controlled incoming traffic rate to a router
[333]	To manage the network traffic problem in migrating parallel web crawler	Mamdani FIS	–	(i) Communication, (ii) bandwidth, and (iii) noise	Network traffic

(continued)

Table A.11 (continued)

Refs	Aim	Fuzzy technique	Combine technique	Input	Output
[334]	To find an optimal routing path in 6LoWPAN networks avoiding congestion	Fuzzy logic	Weights sum model (multi–attribute decision–making approach)	Buffer occupancy, expected transmission count and routing metric of the network	The best parent–node
[335]	Quality of service guaranteed	Mamdani FIS	–	Source packet delivery ration, packet QoS type, queue length and packet expiry time	Packet's priority index

* The reference numbers concern the references in Chap. 3

Table A.12 Examples of fuzzy logic use in network security

Refs	Aim	Network	Security practice/ policy	Fuzzy technique	Combined technique
[336]	Threat detection	Blockchain–based IoT environments	Intrusion detection system	A novel fuzzy DL model, optimized adaptive neuro–fuzzy inference system (ANFIS)–based attack detection, fuzzy matching (FM), and fuzzy control system (FCS) for	Metaheuristic algorithms
[337]	Anomaly detection	Internet of Medical Things infrastructure	Intrusion detection system	Fuzzy C–means clustering	Bi–LSTM technique
[338]	Prediction of security index	Vehicular ad hoc network (VANET)	Prediction model	ANFIS	–
[340]	Improving security and stability	Vehicular ad hoc network (VANET)	Routing protocol	Fuzzy neural network	–
[341]	To ensure the correctness of the information received from authorized vehicles	Vehicular ad hoc network (VANET)	Secure trust model	Mamdani FIS	–
[343]	To enhance network security	All networks	Intrusion detection system	ANFIS	Feature selection algorithms
[344]	To deal with hole attacks	Opportunistic networks	Security protocol	Mamdani FIS	Game theory
[345]	Detection of malicious activities	Internet of things	Intrusion detection system	Fuzzy decision trees	Bayesian possibilistic clustering
[347]	To detect phishing websites	Fog networks	Anti–phishing software	Neuro–fuzzy network	–
[348]	To secure from from jamming attacks	Wireless sensor network	Intrusion detection system	Takagi–Sugeno FIS and ANFIS	–
[349]	Insider attack detection	Wireless multimedia sensor networks	Intrusion detection system	Mamdani FIS	–
[350]	To detect black hole attack	Mobile Adhoc network (MANET)	Black hole detection system	Fuzzy logic algorithm	–
[351]	Malicious node detection	Wireless multimedia sensor networks	Malicious detection system	Mamdani FIS	Improved elliptic curve cryptography algorithm

(continued)

Table A.12 (continued)

Refs	Aim	Network	Security practice/policy	Fuzzy technique	Combined technique
[352]	Detecting sybil attacks	Vehicular ad hoc network (VANET)	Intrusion detection system	Two Mamdani FIS	Arithmetic optimization algorithm
[354]	Intrusion detection	Internet of things	Intrusion detection system	Mamdani FIS	Correlation–based feature selection
[356]	To improve the information security control analysis	Cloud computing networks and wireless sensor networks	Information security control	Fuzzy analytical hierarchy process (FAHP)	–

* The reference numbers concern the references in Chap. 3

Table A.13 Examples of fuzzy logic use in computer security other than networks

Refs	Aim	Environment	Security practice/policy	Fuzzy technique	Combined Technique
[339]	Risk assessment of improper activities	Cybersecurity	Risk assessment model	Combination of three mamdani–type FIS	–
[342]	To detect malware	Android operating system	Malware protection	Mamdani–type FIS	Machine learning techniques
[346]	To manage trust	Mobile edge computing	Trust management framework	Mamdani–type FIS	–
[353]	To evaluate trustworthiness of a cloud service provider	Cloud computing	Trust evaluator system	Four mamdani–type FIS	–
[354]	Detection of not normal user behavior	Computer security–user behavior	User–level intrusion detection system	Three mamdani–type FIS	Finite automaton
[357]	To optimize intrusion detection	Data protection	Intrusion detection system	fuzzy based learning model	Machine learning
[358]	Threat detection	Cisco adaptive security appliance 5500 series firewalls	Threat detection system	Mamdani–type FIS	–
[359]	To ensure data privacy and quality	Cloud computing/smart cities	Anomaly detection system	Fuzzy C–means	Homomorphic encryption and MapReduce model
[360]	To improve the performance of intrusion detection system	Computer security	Intrusion detection system	Fuzzy C–means	K nearest neighbor algorithm
[361]	To classify anomaly activities	Computer security	Intrusion detection system	Fuzzy kernel robust clustering	–
[362]	To detect anomalous behaviors	Data centres	Anomaly detection system	An evolving fuzzy–rule–based classification framework	–

* The reference numbers concern the references in Chap. 3

Table A.14 Examples of fuzzy logic–based navigation software systems

References	Aim	Navigation system	Fuzzy technique	Combined technique
[367]	To detect the maneuvering level of the vehicle	Low–cost SINS/GPS	Mamdani FIS	–
[368]	Performance evaluation of fusion algorithms for integrated navigation system	Navigation systems in general	fuzzy LLSM and FCE method	–
[369]	Improvement of the navigation accuracy of low–cost SINS/GNSS navigation system	Low–cost SINS/GNSS	Type–2 fuzzy sets	Hammerstein neural network
[370]	Improvement of the position speed and accuracy in high–speed vehicular navigation systems	Low–cost INS/GPS	Type–2 fuzzy logic system	Extreme learning machine (feedforward neural networks)
[371]	To guide an automated Guided Vehicle moving along the road marking lines	Navigation system of automated Guided Vehicle	Mamdani–based fuzzy control algorithm	–
[372]	To navigate and control an autonomous underwater vehicle into the docking station in a strategic manner and execute different maneuvers	Navigation system of autonomous underwater vehicle	Sugeno FIS	–
[373]	To detect and compensate abnormal USBL information and maximize USBL information	SINS/USBL	ANFIS	Bayesian kalman filter and principal component analysis
[374]	To determine the current positions and predict possible errors	Indoor navigation systems	Neuro–fuzzy System	–
[375]	To enhance the robustness and the estimation accuracy of a visual inertial odometry navigation system	Visual inertial odometry navigation system	Sugeno FIS	Kalman filtering
[376]	To identify the movement state and type	Loosely–coupled navigation system	Mamdani FIS	–
[377]	To automatically adjust the system's noise covariance matrix, enhancing the positioning accuracy and reliability	Low–cost SINS/ GNSS	Type–2 fuzzy logic system	–
[378]	To improve processing time and enhance accuracy of the prediction of the vehicle location	Low–cost INS/GPS	ANFIS	Subtractive clustering algorithm

(continued)

Table A.14 (continued)

References	Aim	Navigation system	Fuzzy technique	Combined technique
[379]	To define the reward of the vehicle depending on its current state	Navigation system of autonomous vehicles	Mamdani FIS	Deep learning
[384]	To adjust the weighting factor of the process noise covariance matrix of the kalman filter	INS/GPS	Sugeno FIS	Kalman filtering
[381]	To decide about the existence or not of a roundabout	Navigation system of autonomous vehicles	Mamdani FIS	–
[382]	Recognition and tracking of autonomous underwater vehicles	Navigation system of autonomous underwater vehicles	Tsukamoto fuzzy model	Image–color recognition and kalman filtering
[383]	Ship navigation enhancement considering wind, current and waves	Navigation system of ships	ANFIS	–
[384]	Ship steering control	Navigation system of ships	Mamdani FIS	–
[385]	Autonomous vehicle navigation based on RFID sensors	Navigation system of autonomous vehicles	Mamdani FIS	–
[386]	Improving accuracy in mobile robot navigation and obstacle avoidance	Robot navigation system	Knowledge–based neural fuzzy controller	–
[387]	Optimization of the mobile robot navigation	Robot navigation system	ANFIS	Particle swarm optimization
[388]	Steering control of a mobile robot	Robot navigation system	ANFIS	–

* The reference numbers concern the references in Chap. 3

Table A.15 Examples of fuzzy logic–based HVAC systems

Refs	Aim	System	Fuzzy technique	Input	Output	Combined technique
[389]	Residential energy management	Thermostat in heating/cooling systems	Mamdani FIS	Environmental conditions, house parameters and characteristics, user preferences and schedules	Set point of the temperature	–
[390]	To forecast meteorological data for thermal systems	Thermal systems	ANFIS	Months, number of sky clear days, cloudiness, average air pressure, ground surface minimum temperature, mean solar time, and evaporation	Temperature, solar radiation, and relative humidity	Back–propagation network based artificial neural network
[391]	To increasing thermal comfort and reducing energy consumption	Air conditioning thermal controller	Mamdani FIS	The indoor air temperature and the occupants' actual thermal sensation	The set point temperature and the fan speed	–
[392]	To achieve low computational complexity while improving the occupants' thermal comfort	Smart thermostats	Two Mamdani FIS	1st FIS: the weather data, the available funds and the current market's energy trading rate 2nd FIS: the occupants' thermal comfort and the thermostat's temperature set–point	1st FIS: availability of energy 2nd FIS: current demand for energy consumption	Artificial neural network
[393]	Maintenance of thermal comfort and energy saving	Smart thermostat in air conditioner	Mamdani FIS	Electricity prices, solar radiation, and occupant presence	Set point temperature change	–
[394]	To reduce energy consumption	Smart thermostat in air conditioner	Mamdani FIS	The internal ambient temperature of the neighborhood offices, the internal ambient temperature of the open space, the external ambient temperature	The new setpoint temperature	–
[395]	Temperature and humidity control	PID controller of HVAC system	A decoupling fuzzy–neural network	Environmental parameters	Temperature and humidity setpoints	–
[396]	Fault detection with limited faulty samples	HVAC	ANFIS	Real–time monitoring data of the building energy system	Fault existence or not	–
[397]	Energy consumption reduction	Air–conditioning system	Mamdani FIS	Set point temperature, outdoor temperature, physical presence	A multiplying factor for the set point temperature	–

(continued)

Table A.15 (continued)

Refs	Aim	System	Fuzzy technique	Input	Output	Combined technique
[398]	Automatic and dynamic control of the speed of the HVAC evaporator fan	HVAC	Mamdani FIS	The occupant's desired set point temperature, the temperature difference between the actual space temperature and the desired temperature	The speed of the evaporator's fan	–
[399]	To maintain the indoor thermal comfort	Air conditioner	Mamdani FIS	The predicted mean vote index, the variation of the predicted mean vote index, the outside temperature	The fan speed	–
[400]	Adaptive energy management in commercial facilities	HVAC	Mamdani FIS	The current temperature and the number of consumers	The heating and cooling functions	–
[401]	Thermal comfort provision	HVAC (heating controller, cooler controller, window controller)	3 zero–order Sugeno type FIS	1st and 2nd: the error value and the error change 3rd: the predicted mean vote index and the outdoor temperature	1st and 2nd: the incremental change in the flow regulator position 3rd: the window opening fraction	–
[402]	Energy consumption reduction	Cooling system	Mamdani FIS	Temperature and airflow	Shutter. fan, AC	Backpropagation neural network
[403]	Efficient energy management	Heating and cooling system	Mamdani FIS	The difference between the actual and the desires temperature, and the rate of temperature change	The operating modes of heating and cooling system, and the fan rotation speed	–
[404]	Fault diagnosis	HVAC	Knowledge–driven fuzzy inference system	Room temperature, daily temperature, CO2 concentration, occupancy	Fault category	Bayesian belief network
[405]	Energy saving and cost reduction	HVAC	Mamdani FIS	Error and change of error	The variation between the indoor air temperature and the set–point temperature	Particle swarm optimization algorithm
[406]	Automatically adjustment of indoor climate conditions	HVAC	Mamdani FIS	Derivative temperature, error, indoor–outdoor difference, humidity	The temperature set–point	Random forest algorithm
[407]	Minimize energy consumption	HVAC	Mamdani FIS	User presence, demand, outdoor temperature, and price	Setpoint temperature	–

(continued)

Table A.15 (continued)

Refs	Aim	System	Fuzzy technique	Input	Output	Combined technique
[408]	To maintain the room temperature at a desired level	Air conditioner	Mamdani FIS	Temperature error or difference between the current room temperature and desired value, integral error, differential error	Signal for cooling or heating	–
[409]	Occupant comfort levels improvement	Central air conditioner	Interval type–2 fuzzy logic system	The occupancy rate change of building's zone	The weights for energy and comfort optimization	A non–dominated sorting genetic algorithm
[410]	Adjustment of the thermal comfort and indoor air quality in a vehicle's cabin	Vehicle HVAC	Mamdani FIS	The predicted mean vote index, the carbon dioxide concentration, and cabin temperature	The blower velocity, the mass flow rate of the refrigerant and hot coolant, the recirculation flap, and the HVAC mode	–
[411]	To control the air flow and quality	Ventilation system	Mamdani FIS	The indoor temperature, the humidity, the air flow, and the microorganisms in air	The fan speed, the cold air, and the operation mode	–

* The reference numbers concern the references in Chap. 3